PESARO ARCHITECTURAL MONOGRAPHS
Editor – Patrick Bingham-Hall
Associate Editor – Philip Goad
Design – Kirk Palmer Design
Printer – South China Printing, Hong Kong

Pesaro Publishing
PO Box 74
Balmain NSW 2041
Australia
Phone 61 2 9555 7422
Fax 61 2 9818 6999
Email bingham@hotkey.net.au

PESARO PUBLISHING

First published in 1999 by
Pesaro Publishing
Sydney, Australia

ISBN 0 9577560 0 3
ISSN 1442-8393
Margalit, Harry Zwi
Durbach Block: the luminous space of abstraction
Bibliography
1. Durbach Block Architects
2. Architectural firms - New South Wales
3. Architecture - New South Wales
4. Architecture - South Africa
I. Goad, Philip
II. Bingham-Hall. Patrick
III. Title. (Series: Pesaro Architectural Monographs; 1)
720.9944

1

DURBACH BLOCK

THE LUMINOUS SPACE OF ABSTRACTION

CONTENTS

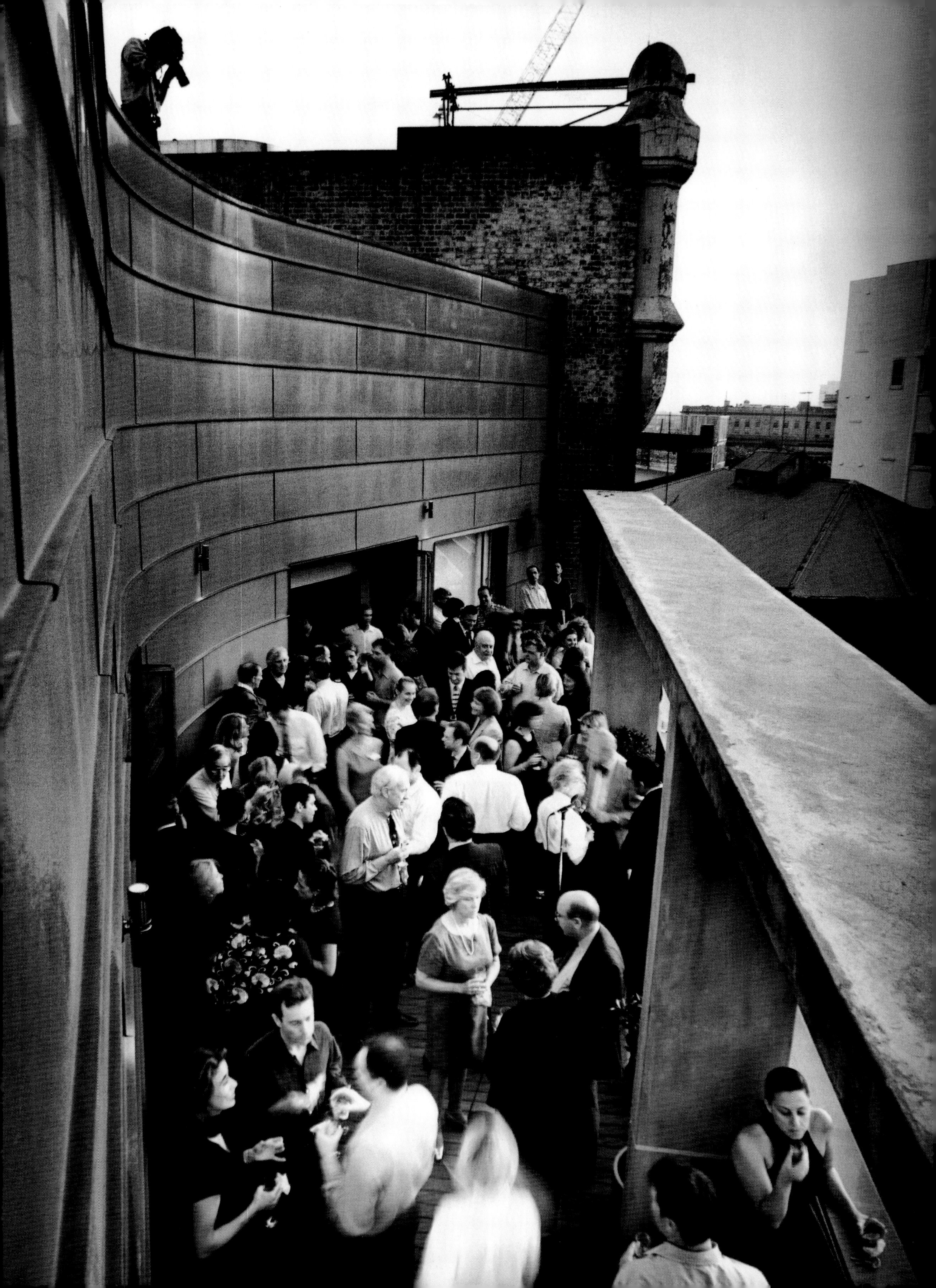

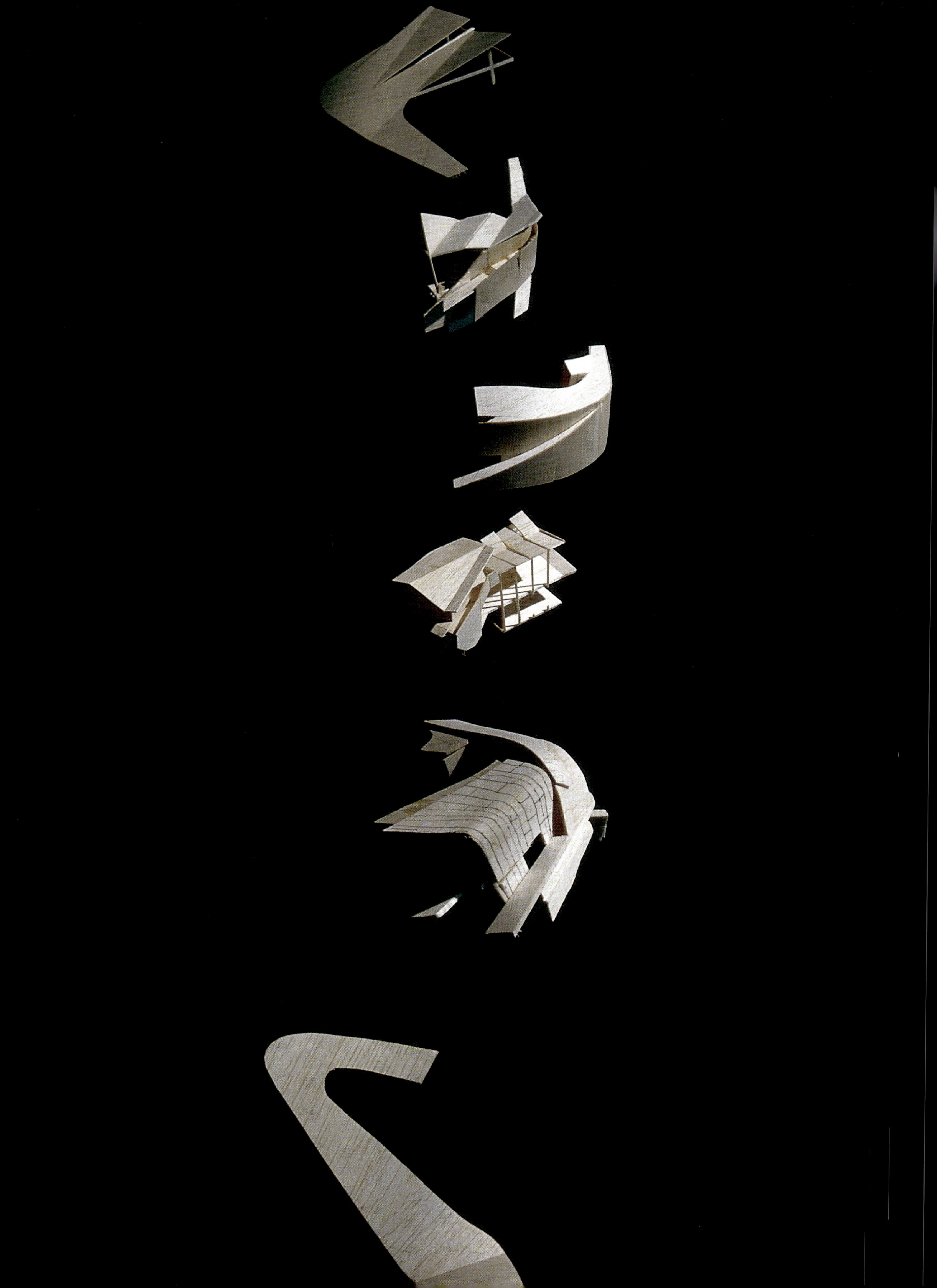

INTRODUCTION

STANLEY SAITOWITZ

Of the trajectories which have emerged as the field of contemporary modernism, the legacy of Mies and hyperminimilism form one broadly explored pole, with the more complex and currently less traversed field of Le Corbusier as the other. It is this other that is the fertile incubator of the Durbach Block work.

From this tradition, the cartography of plans generate both geological and choreographic fields. They re-map their found landscapes, incising paths which weave new topographies of procession, staging haptic worlds of volume, structure and light, furnished with the machines of function.

From this tradition is the courage for invention over convention, the preference for prototype over type, the reliance on truth of imagination over reason, and the choice of poetry over philosophy. Good taste, or any taste really, is hardly a subject in the process of creation which deposits these powerful new instruments in the world.

And as the work develops through the pages of this book and time, matter becomes ever more sublime, diminishing the role of form and expanding the realm of space.

A SHIFTING SENSE OF BEAUTY

HARRY MARGALIT

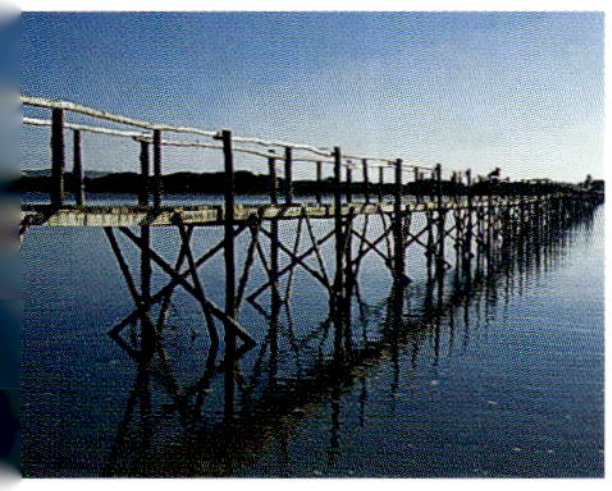

The work of Durbach Block sits outside of many of the prevailing obsessions in Sydney architecture. As Sydney comes to terms with its good fortune in becoming the object of considerable interest and mythologising in the international architectural press, Durbach Block skirt the quest for a definitive regionalism in favour of a less prescriptive way of working. Striving for an immediacy of response, the layering within their work nonetheless speaks of the concerns of personal migrations, with their own cadences and aspirations. The work in this volume covers most of the years Neil Durbach has spent in Sydney. Camilla Block's history is somewhat different, although with a similar geographic origin. It would be a mistake to say the same origin. Camilla spent her early years in Johannesburg, Neil in Cape Town, the two cities representing a polarity which Pevsner, many years ago now, described as being further apart than any two cities in Australia.

It is a truism that South Africa is a deeply historical place, in the sense that history breathes down its neck like a jockey over a blood-flecked racehorse. In his early years Neil thrived on this with a particular temperament which allowed him to harness its potential while avoiding its hazards. For Cape Town in the 1970s represented an amalgam of break-neck history with an overlay of culture, the two quilted together in an often surprisingly sophisticated manner. At hand were traces of a range of modes of production, to use a term of wide currency at the time. Before the dismembering of Cape Town under apartheid there had existed one of the most Kipling-esque of cities, somewhere really of the east on which were sustained parts of the occidental dream. This variegation, initially structured around the quarters of a British imperial city and later partially preserved and partially destroyed under the pax Afrikaner, meant that it was a city holding traces of slavery and indenture alongside a broad-chested modernity. Materials themselves retained these associations: the solid masonry of Dutch colonial buildings gave way to the more precise rendered surfaces of artisans trained in the English techniques, but it was always shot through with a sense of many hands whose imprints were left in the solidity of its construction.

It is hard to forget this solidity, even in the face of the scarcity of working hands which marks much of Australia's recent history and makes for a receptiveness towards technology and a quickness of doing. Rendered, whitewashed, old Cape Town somehow allowed a quick mental calculation of lives lived, beauty bought, blood spilt and debts owed. And it was this last, the debt affluence owed to poverty, which underlay much of the intellectual vigour, nay arrogance, which Cape Town sported. The counterculture of the 1960s found fertile ground there, and not primarily as a generational schism. The early sixties was the time when apartheid hammered down its intent. Liberalism, for many affluent English-speaking South Africans the politics of choice, reinvented itself as a sort of pressure-equalisation mechanism of the psyche. Its vision of a guiltless equality under law was sublimated into rivulets of idealism, their flows and gatherings occasionally coloured with the intensities which the situation engendered. The substance on which this idealism acted was drawn from the range of countercultures

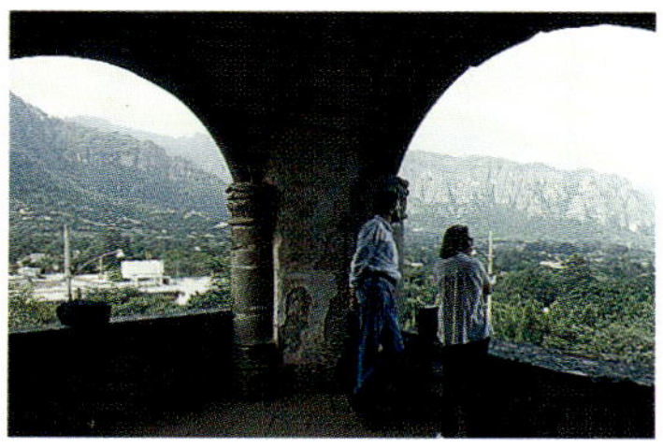

both American and European, and architecture was no exception.

One such glittering strand was the work of Roelof Uytenbogaardt, then Professor of Planning at the University of Cape Town. Although having little to do with the School of Architecture, he was preeminent as the author of intellectually uncompromising buildings. In the early 1970s Roelof was practicing an architecture of steadfast abstraction, unconcerned with sentiment or the vernacular picturesque. Despite his work standing in sharp contrast to the Cape vernacular revival well underway by then (in exactly the mediated sense which Frampton championed), it was widely supported by the profession precisely because it offered an alternative, trenchantly progressive vision of architecture's redemptive possibilities. In such a cleft society, where else might we all meet but in the luminous space of abstraction? This is, of course, the prescription for a modernism of ideas, which trumps obsessions with materiality and technique.

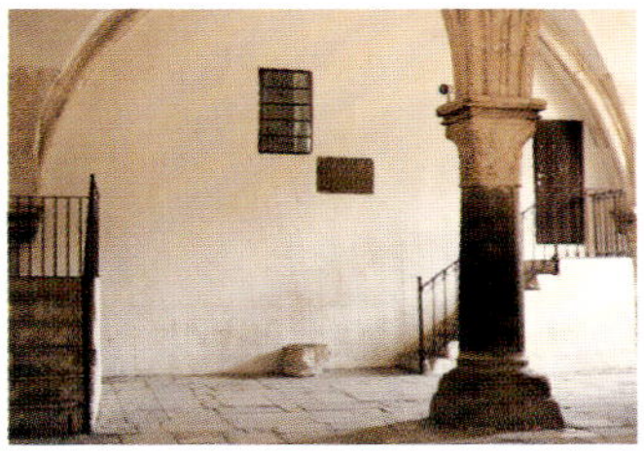

This unswerving commitment to the idea of the building has persisted in the work of the Durbach Block. It is not conceptual architecture, but an architecture of concepts where the realisation of the project is its most exciting consummation. Neil has carried the images of Roelof's work around with him (as have many Cape Town graduates from the past 25 years) as a reminder of this primacy of thought, of the pitch to which the high modernists drove their conceptual tools in the fashioning of their works. The case of Roelof is instructive because the school of architecture in Cape Town in the 1970s involved few outside practitioners. It was an academic school which, curiously, taught little, relying on the cultural affectations of students to carry them through as they set their own agendas. Thus the few examples of truly uncompromising architecture which Roelof and a handful of others built carried enormous influence through their disdain for prettiness.

Johannesburg at the time had a more eclectic, but more prescriptive, architectural culture. The energy of Pancho Guedes, having recently quit Mozambique, was palpable, and the Iberian twist he gave to aspects of the vernacular encouraged in the work at the University of the Witwatersrand a formal experimentation which Cape Town was too involved in abstraction, too cool to attempt. Neil nonetheless sought out students and young architects from the north, from Johannesburg, precisely because they were demonstrating the possibilities of form, working in ways which were concrete, evocative, playful but above all productive. Their very concreteness was a liberating force, a means to free the hand from the stern dictates of intellectualism. Some influential figures from those years have gone on to prominence, none more so than Stan Saitowitz. Their contribution to Durbach Block, both via friendships and via the couple of years Camilla spent in Johannesburg as a student, is manifest to any eyes familiar with the forms of Pancho's work, or Stan Saitowitz's Transvaal House.

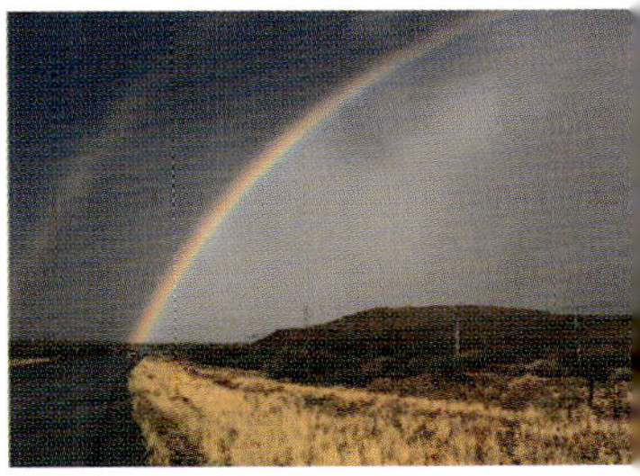

Perhaps the main contribution of these works to the Durbach Block ethic is that they engendered a dialectic (no other term seems quite as appropriate) which oscillates between a love of the abstract possibilities of high modernism, and the critique of its limits through a parallel fascination with form and materials. No practice can grow without an essential tension: it drives the work beyond a set language, and demands of each project a new, momentary resolution of the tension. Thus Durbach Block are committed to modernism, yet their work is an ongoing investigation of its shortcomings. They are committed to beauty, yet deeply distrustful of prettiness.

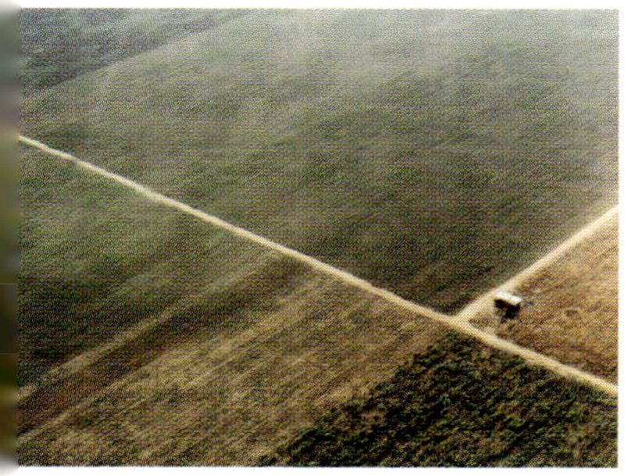

This confluence of concerns is more fortuitous given the gap of half a generation or so between Neil and Camilla. Coming to Australia as a child, Camilla nonetheless remained conscious of that distinctive South African milieu which encourages a relentless idealism in all things. Australia tempered this with its sense that not everything is broken. Sydney University School of Architecture also claims the right of each individual to set their own agenda, and Neil's teaching stint there as a part-timer in the late 1980s led to many enduring bonds, including the association with Camilla and Nicholas Murcutt He was an unusual teacher for the time, demanding of students not that they construct things from first principles, but that they demonstrate some spirit in their work as well as an ability to discern between good architectural ideas and bad. This approach derived from a confidence in his own understanding of architecture, which is broad but unapologetic.

His forthrightness, coupled with what might be described as pure synaptic talent, was always going to lead to something. Within months of arriving in Sydney Neil had some success in a competition for the redevelopment of Circular Quay. His partner in the winning entry was fellow Capetonian Harry Levine, an accomplished exponent of the tough idea. Immigration demands of skills the same as it demands of furniture and clothes – that they can be packed down, folded and transported, hopefully to be unpacked and sustaining on the other side. The Circular Quay win gave some hint of effective intellectual packing, and the pair's success a couple of years later in the competition for the extensions to Tusculum for the Institute of Architects was testimony both to the clarity of their reading of the problem as well as to a certain iron professionalism in the presentation of the solution. The partnership dissolved after what proved a tough commission, Neil nonetheless persisting with the aim of maintaining an independent practice.

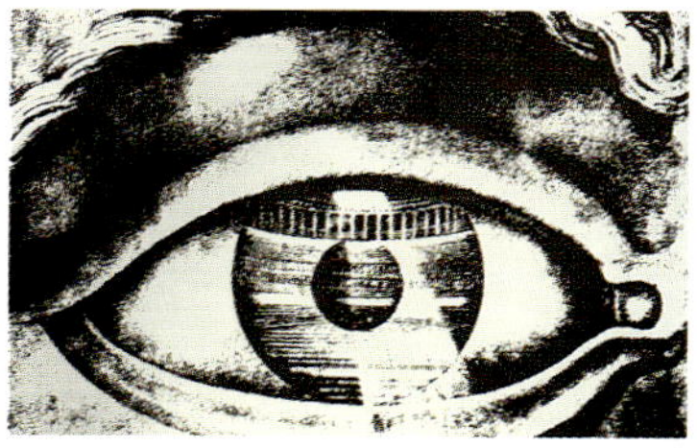

Tusculum, for all the difficulties it presented, demonstrated that aspect of Sydney where the city will accept your credentials if it believes you can do the job. Its not that you will be spared criticism and suspicion, simply that you will be subjected to the same criticism and suspicion as everyone else. Ostensibly so similar to white South Africa with its boats and trailers and outdoor paraphernalia, Sydney has an entirely different set of longings. In architecture, its moments of transcendence hover between building technique and a new world mysticism. Here ideals can be at once intensely practical and persistently elusive. These poles manifest in a strong tradition of simple yet effective detailing on the one hand, and on the other a tradition of Heideggerian metaphysics which mirror a restlessness of identity. This is fertile ground for regionalist or nationalist architectures, and they have indeed flourished. But it is very different from the Cape Town tension between history and redemption, and its loose regard for the practical.

For Camilla, the synthesis of two identities has given her a strong sense of architecture as a shepherding of ideas through the exigencies of the building process. She is accomplished in the traditional architectural sense, and her willingness to take the necessary road in order to make manifest the integrity of a scheme has put some steel into the firm's way of working. The current modus operandi is, predictably, different from the early years. House Vinh, for example, was a strange, rolling affair of the late 1980s and early 90s, with a client of unpredictable and occasionally startling generosity who spent and spent and one day decided to stop spending. His largesse, as far as it went, allowed the remodeling of a large but undistinguished house with a degree of experimentation which could not have happened any other way. There was something almost medieval about the process, with the architect as artist and intellectual jester, but the lessons of the house were extended in a number of inner city renovations which were discrete exercises in the crafting of space.

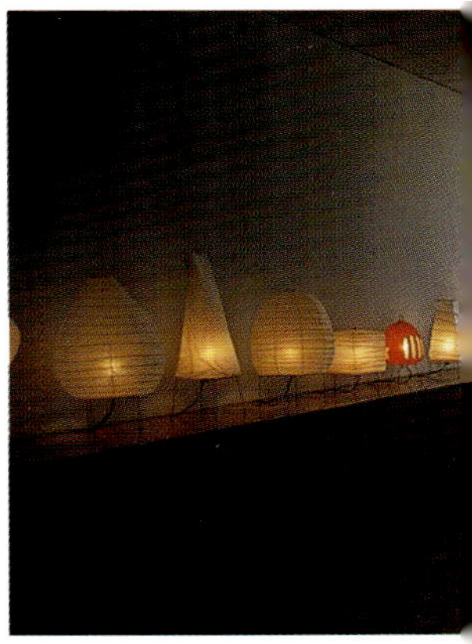

The methods of Australian building are distinctly different to those in South Africa, where masonry prevailed. The framed techniques of Australia, as in North America, allow great flexibility and speed of construction, but it takes time to absorb their possibilities. In House Vinh much of the experimentation was around this point: a teasing of the terrific framing skills of Australian carpenters. Other substantial renovations pressed the point, as their spaces were sheeted and cranked and elongated. The constraints of renovation were a fruitful area for experimentation and honing of skills. Any existing building, if it is to be retained, demands a tectonic reading of sorts, a discerning of form and possibility which forces the architect to work towards a solution which is somehow comprehensible. A list of projects from the early years of the firm testify to its involvement in renovating and extending. The trace of this period and its lessons are the models which lace the office, and which form a crucial part of its method. The models serve to provoke discussion about form and space in precise terms. Through them these ideas of comprehensibility, of sequence and teasing and rest are developed. The firm is more aware than most of the constant translation which occurs, albeit fairly transparently, between drawings and space. Developmental models circumvent this to some degree, and through models spaces are sketched and honed. One consequence of this has been the firm's straightforward transition from drawing boards to computers. Having never been reliant on a series of graphic transformations to develop a design, they continue to use models and sketching to formulate a precise intent, which is then quantified onscreen.

The Darling Point house has taken on a singular significance in the firm's history for a couple of reasons. It seemed to mark a transition from a predominance of renovations to a wider range of built work. It also held distinct technical challenges, where framing and lightweight cladding were used to overcome the problems of access to the site. These were extreme to the point where the holes for the piled footings were augered by hand, the vision of a Popeye-like set of forearms wrenching the device down to rock

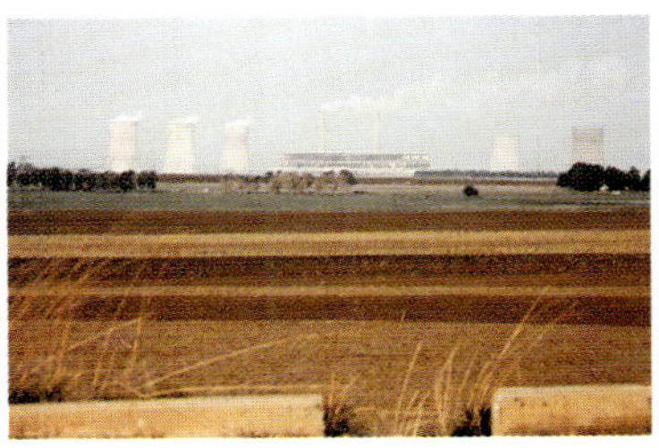

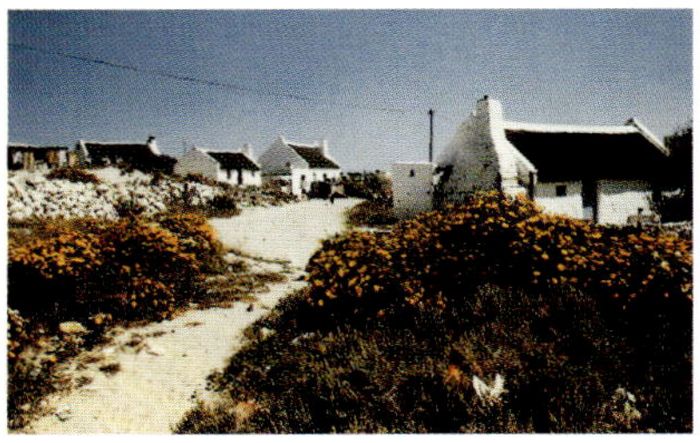

forever etched into the collective memory of the office. The house itself reveals something of a schism – its interior, faceted and airy, could only have been made out of lightweight materials, but the exterior mimics masonry. Here on the outside something of the impermanence of the 1980s Sydney vernacular might have done better, given that the whole experience of the building is vertical, culminating in the roof terrace and improbable pool and bridge. But the painted surface seems to have been favored because of its potential for renewal, in the same way as a masonry architecture, rendered and whitewashed, can experience repainting almost as a festival which keeps history at bay.

The growing familiarity with a greater range of building techniques and problems paid architectural dividends. In subsequent years the materials began to inhabit the intellectual formwork more comfortably. With the Hillside House the whole process seemed to have settled: concrete and timber, steel and brick, each lent its required gravitas or delicacy as needed. Ostensibly the buildings, in conception, still displayed the clarity of intent and crafting of space which had been there at the beginning. The method endured, too, with analysis laid over a careful reading of context and circumstance in a process which always has one eye turned to the obvious. For the obvious, the everyday, is a referent which guarantees utility, as well as ensuring a connection between architect and inhabitant. Both Neil and Camilla are too canny to ignore this: much of the work of the practice is directed towards showing the everyday in a different light.

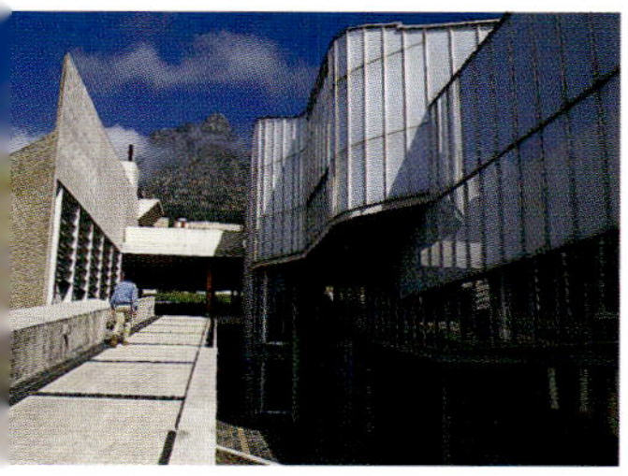

This clearly places them on the side of the artist provocateurs. The historic irony here is that they do it with a degree of consideration which underscores a view of architecture as a kind of seduction. The newness of each project is constantly checked against an aesthetic sense which has been developed precisely to act as scaffolding and support, as the work is cantilevered out into the public realm.

This method is deliberately tested in design competitions. The datum, the starting point, is generally the formulation of the most obvious solution to the demands of siting and program. Then, in a sort of scanning motion, site and brief are scoured for alternate starting points. Taste, prejudice, historical resonance: all these generate schemes and fragments which are then tested against the initial solution. Most are discarded, but some pass this evaluation surprisingly well, throwing off unexpected benefits. These ideas are gathered and refined, made comprehensible and given nicknames. They are then scrutinised for potential, held one against the other, and a decision is made to pursue one or two. The process is often laboured, and sometimes circuitous, but it has the virtue of testing ideas at some length to find the ones with legs. Above all they must have a clarity, which often emanates from a contextual order extending beyond the confines of the site. The schemes for Wagga Wagga and Federation Square illustrate this point as well as any: the prevailing urban logic which they reinforce is counterpointed by the singular aesthetic which links together the various parts of the proposals. These schemes hang intentionally between desire and utility, between what is and what might be. Curiously for a practice with strong roots in a highly politicised milieu, the work carries over more of a sentiment than a politics, more of a striving than a dogma.

In the case of two recent built works, the Foster Street apartment and the Olympic amenities blocks, these concerns manifest in more mediated ways. Here the lessons of Claes Oldenburg resonate, especially the idea of an urban comprehensibility which relies on flashes of scale and gesture, on objects whose worked forms leave an afterimage. The Foster Street apartment, internally, thrusts its inhabitants into Sydney's skyline. From a distance, though, its curve and cladding present a beaten metallic piece to an urban quarter still dominated by the hand-wrought technologies of the turn of the century. The Olympic blocks, too, take their cue from context, in this case the cleared industrial land which forms the Olympic site. If any sentiment dominates the surrounding buildings for the games it is an earnest sobriety, a collective demonstration of how to spend building budgets responsibly, with a measure of ingenuity and some evident sense of natural design. What the amenities blocks bring is an unabashed funkiness, unique in the site. This contribution, as contextual as any, also provides luminous beacons like great cats' eyes to render distinctive certain parts of a vast and hurried development.

Even a firm which carries much of high modernism's aspirations needs to acknowledge, now and then, the irony of such a position. Despite a clear commitment to the tough work of developing form which has a sense of currency and a sense of the future, the work of Durbach Block also represents a particular critique of modernism. A close look at many of the most crafted spaces – Einfeld, Kwok and the Hillside house, for example – reveals a deliberate blurring of the line between a utilitarian logic and one streaked with a certain wilfulness. The intent, slightly sardonic, is to maintain an edginess, a restatement of the role of modernism not as the triumph of common sense but as a practice which prods one, by varying degrees, beyond the familiar.

This distinctive edginess, this toying with the logic of architecture, brings us back to the question of an Australian-ness. If there is any identity proposed here, it lies in accepting that there is no architectural closure, no homecoming. If skills can be packed down and unpacked, this process invariably effects a transformation on its components. Somehow in the unpacking the fleshiness of a particular formal language disappears, leaving the bones and sinew of abstract fascination. This changeable curiosity, brought to bear on the process of design and driven by a restlessness towards solutions even as they occur, leaves behind a trail of buildings which have tested different constellations of client and circumstance for their possibilities. Seen in this way, the buildings and projects reiterate a sense of the ever-in-the-present, a shout in the wind occasionally softened by a meticulous but shifting sense of beauty. All that is required when viewing the work is not to hide. Perhaps aesthetics is the technique of overcoming the perpetual unfamiliarity of the present: the work of Durbach Block rests on us admitting this as a possibility, and then being engaged, perhaps delighted, by its ramifications.

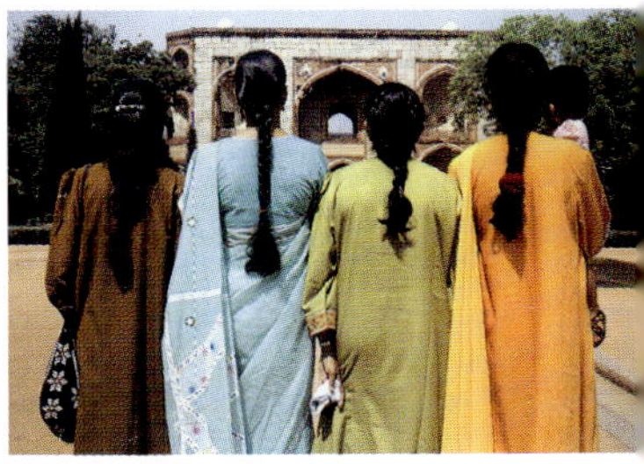

LIVING WALLS: 13 PROJECTS BY DURBACH BLOCK

PHILIP GOAD

In a Beaux Arts plan, the term poché refers to the blackened-in areas which delineate the boundaries of walls. Poché is black. It suggests mass but it is largely atectonic – one doesn't really know how the building is constructed or pieced together. Poché indicates the position of walls that bound space but it often gives no indication of what happens in the black space – what happens between.

To begin a discussion of the work of Sydney architecture firm Durbach Block by invoking the drawing techniques of the Ecole des Beaux Arts is dangerous. One risks incurring the subjects' wrath, their squirms of discomfort at an implied association with orthodox (and decidedly unfashionable) academic strategies. Durbach Block would refute any such connection. But consider the strategy of the obverse of poché – that instead of it defining space external to itself, that a free poché is itself the primary container of fully bounded space. The thirteen projects of Durbach Block are about the spaces between. It is an architecture of plastic poché, a geomorphic architecture of living walls. Each project is characterised by inhabited walls that are bent, skewed, canted or carved into. The idea of wall is not employed to create an isolated object. Instead, two strategies inform these thirteen projects: architecture as urban wall; and architecture as landscape wall. This body of work therefore is not an architecture of heroic villas or urban monuments. There is an obligation to a larger picture, a level of humility in the face of a broader scheme of figure-ground that comprises the city and the natural landscape.

ARCHITECTURE AS URBAN WALL

A length of wall can frame an urban space, complete it. A series of wall lengths can define slivers of spaces such as streets, lanes or corridors. A wall length can be bent to form a court, to contain space. Each wall length has habitable thickness. This is architecture as urban wall. A wall length may also have plasticity which confounds a simple reading of it as a man-made form. This malleability means that analogies for such walls may even oscillate between city and landscape. There is an intriguing blur between the two.

The RAIA (NSW Chapter) Headquarters was a discrete addition to John Verge's Tusculum (1830-32, 1851) at Potts Point. It was the creation of a backdrop 'wall' to frame a court before the restored Italianate (once Regency) villa. The outside faces of this 'wall' can be scaled by numerous stairs, while inside, this massive block was hollowed out to form a foyer, offices and theatrette. In detail, there are personal debts to Le Corbusier and Carlo Scarpa, but overall there is a newcomer's deference to Verge's formalism and the Sydney penchant for weathered stucco so ardently loved by diverse practitioners like Leslie Wilkinson, F Glynn Gilling and more recently Espie Dods. This early project was polite and recessive as it bowed to a venerable Sydney icon. In later projects, the idea of urban wall was taken as a compositional tool in itself. It could be used to create its own city.

The urban wall has also taken the form of an inserted fragment. In the Foster Street building, the addition of an penthouse apartment was a craggy outcrop clad in zinc. A concrete brise-soleil was the only concession to architectonic form, a framed view amidst a gritty sub-landscape of tall brick warehouses. In the much earlier Vinh house, the strategy was the reverse. A chasm of space – a plaster ravine for an entry hall with a crafted timber ceiling way above – was inserted within an existing mass of house.

In the shortlisted competition entry for the Wagga Wagga Civic Centre, great thick walls of building denote three massive fragments (one of which is an existing 1950s theatre) which, when placed together like giant incised pebbles, denote an urban space framing the town's billabong lagoon and opening up a spray of vistas to the Murrumbidgee River. Like Le Corbusier's General Assembly at Chandigarh (1953-61), each of these buildings harbours an honorific space: a council chamber, a civic theatre, and a reading room. Terraced pools and gardens like cultivated fields are laid out with unidirectional force, and accentuate the implied framing of space and vistas. These living walls denote an angular S and a heavy-based U. Perimeter blocks echo the L-shaped form of Walter Liberty Vernon's courthouse (1900-02) which hugs the street line and shields its precious volume behind.

Federation Square in Melbourne, another competition entry, was to be defined by a massive forecourt to William Butterfield's St Paul's Cathedral (1880-91). Bounding the space to the south is series of glass slivers that slice through a podium. They provide vistas that connect again to a river, this time Melbourne's Yarra, but now made solid as glazed crystal peaks. These are living walls which house wintergardens – a sleight of hand to reverse figure–ground in Melbourne's cool southern clime. To the forecourt's east, the Cinemedia/Performing Arts Centre is made by folding a wall into a helix. Instead of Corbusier's spiralling Mundaneum (1929) which had the highest point (and thus man) at its centre, there is, in Melbourne a simple garden as the natural and grounded focus.

Stage I of the new RMIT University Housing at the Bundoora campus on Melbourne's outer fringe consists of a series of walls. There is a spiral U (the glimpse of another helix) and two I's whose ends just stop short of touching each other. This is a collegiate figure-ground typology that has been designed for a future growth of contained spaces and captured views. These living walls are to be clad in timber, softly monumental and textural, with splayed cedar skirts that lift to show a glazed undercroft.

The most recent urban walls are more like pieces of urban furniture. Strategically located at peripheral locations near car and bus parks, the amenities blocks at Homebush Bay are virtual renegade structures. Instead of brazenly baring their structure like the nearby stadia, these eccentric walls have a fabric skin pulled tightly over their bony structure, almost as if prudishly wishing not to expose the truth of what lies within, or else simply to tease. The three blocks are roofed by a series of differently sized steel portal frames. The result is not a constant building section – the complete reverse of the repetitive section of almost every other Olympic structure nearby. Instead there is an alluring warped form. Flontex (a German-developed teflon fabric) has been slung over the top and with the eccentrically disposed portals high above, a white light filled volume is created. At night, each block glows like an oversized Isamu Noguchi lamp, a luminous chrysalis or a biomorphic signpost amidst endless bitumen car parks

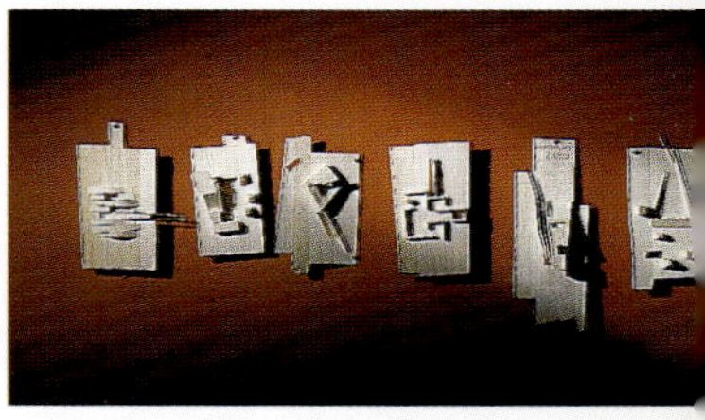

ARCHITECTURE AS LANDSCAPE WALL

Empathy with surrounding natural landforms reveals the idea of architecture as landscape wall. Landscape line and contour are folded into the architecture of the living wall to inform its shape, its profile and above all, its plasticity. In each project, a fluid geomorphic character is pronounced. There is also an admission of architectonic landscape forms: retaining walls, buttresses, paths cut in cliff faces, kerbs, terraces and ramparts.

The Darling Point house comprises two thick living walls around which one climbs, sidles past and bridges, like giant rocks to be circumnavigated. One 'wall' is a boulder shaped slice containing the major living spaces while the other, an irregular trapezoidal bedroom block, seems like a splinter off the stone retaining wall behind. Atop it, there is a framed pergola, the only concession to framed shelter. All else is contained within mass. This house is an outcrop, an eyrie perched on the side of a cliff.

The Hillside house is carved into its cliff with garden courts eroded into its face. Within, steps and passages and enclosed service spaces form the secret paths behind shelf like platforms which open to the view. This house is a landscaped edge. One side of the wall is the earth, the other is a scooped out face with a framed shelter, not for people this time but the car.

The Vista Street house is a liver coloured brick rampart that emerges from its sloping site. It is a massive promontory within a sea of conventional and undistinguished suburban houses. Its folded roofs are like wind-blown trees barely lifting above the rock. Instead of framed shelter, a strutted balcony becomes the only architectonic addition to the major mass while in the back garden, the swimming pool appears like a natural pool caught in the rock's flat top.

Deep within a 1959 apartment block at Darling Point, the Einfeld apartment has been reconfigured, reshaped to reflect the shoreline of Double Bay beyond. The contour and shape of the harbour's edge have been re-inscribed into the shapes of benches, the lines and glass facings of cupboards. The use of differing shades of watery green glass reflects the shifting green depths of the water below. This apartment has had its exterior folded back upon its surface so that each surface becomes its own eroded harbourside cliff face, its own dappled pool of rippling light.

From the living room of the Hunters Hill house you can see across the water to the Gladesville Bridge. The bridge is a deceptively simple piece of engineering – two reinforced concrete arcs of different curvature, a sliver of man-made structure between the sky and the water of the Parramatta River. The roof of the house is a garden. What would normally be the garden at ground level is a pool with a crazy-paved island in it. The house itself is contained entirely within what is essentially a habitable retaining wall, a living landscape wall. The pond is marvellously too large – it is luxuriant – not just for the goldfish but for the site. The pond contends with the house. It gives it reflected light and also makes the site an entire piece of reconfigured landscape – in the tradition of Roberto Burle Marx and Garret Eckbo whose biomorphic hard and soft landscaping were so highly overwrought as to become complete conflations of modernist architecture and landscape. Building and landscape became entire artworks.

The work of Durbach Block has the same precise sense of phenomenological economy as a bridge, a retaining wall, a cliff or a promontory. Their compositions owe much to distant or immediately adjacent views of landscape and constructed lines of urban form. Durbach Block re-orient the picture plane and fold these lines, the sky and the water back into the earth. These projects are different from the prevailing Sydney penchant for sleek value-free asceticism. They are instead a plastic architecture that makes complicated shapes, contains complex spaces and harbours a rich lineage of modernisms – the tropical and sculptural interpretations of late Corbu that emerged from Brazil through Niemeyer and Da Costa, and from South Africa through architects like Issy Benjamin, Derek Crofton and Mannie Feldman. These projects also evoke a modernism whose baroque possibilities afforded by reinforced concrete was exercised by emigre European architects in virtually all Australian capitals in the 1950s This has been largely edited from any historical overview and remains relatively unexplored. In Sydney alone, names such as HP Oser and Hugh

Buhrich suggest a different response to that city's landscape. There is in these projects, the place-making and body-accommodating carved white volumes of Greek island hilltowns so admired by Bernard Rudofsky. Another modernist legacy is the outdoor terrace with its view framed by a concrete beam – the Corbusian tactic of simultaneous containment and release.

At the same time there are the more practical limits of site and context. For example, Hunters Hill in Sydney is full of grand sandstone Victorian houses set in vast gardens with distant views of water and bounded by sandstone walls that bougainvillea spills over. One such house with a late Federation addition of gables, shingles and gingerbread bargeboards sits immediately to the north of the Hunters Hill house. Any new house might have blocked its view and been thwarted at every move in the planning process. Durbach Block were positively generous. They made the Hunters Hill house a giant planter box/retaining wall. The roof was even planted as a garden. The new house disappeared from view entirely and became a visual extension of their neighbour's 19th century verandah.

Inside each of these projects, the theme of the inhabited wall is strongly developed. Walls can house useful spaces over their gradually thickening and contracting length. A stair, a window-seat, a cool built-in concrete chaise-longue (a tribute a la Savoye), a powder room, a wardrobe whose face undulates and, a skylight from which spill stripes of bright white light. Windows with sill heights at lower waist level can screen forgettable houses below and frame trees, water and yet more sky. Incisions with plans can provide spaces for open terraces, sources of light, but also places to open up and enable the most complete erosion of a living wall.

While the language of these projects – mostly white painted abstract concrete forms – could be conceived by many as orthodox modernism reworked, the overall result is something altogether more artful and attuned to the alternate pliability and friability of Sydney's coastal landscape. On the one hand, it appears as if these projects have been laminated to their sites. On the other, it is as if giant wall lengths have been thoroughly eroded, to become percolators of light and air. With either reading, each project loses the self-consciousness of the pristine modernist container. There are not the material and organic analogies of the Griffins at Castlecrag. Nor is there the romance of craft – the 'prospect-shelf/cave' fetish of Sydney's feted placemakers. These projects are not to be regarded as minor siblings to the heroic curves given spectacular speculative venting by Harry Seidler, and they are also not like Steven Holl's intensely formalist but often placeless sculpted objects. Each of these Durbach Block projects instead describes a charmingly old fashioned idea that through a lyrical rereading of a useful structure – a wall, or a bridge, and through a close reading of the landscape – a cliff, an escarpment, or a shoreline – that invention might occur.

This analysis is clearly a formal one. It overlooks the intricacies of biography and attribution, and it does not attempt to penetrate the misty poetic stance of its protagonists. The work speaks for itself. What this architecture indicates is a sensitivity to place, especially an intimate understanding of the Sydney landscape that moves beyond any other previously proposed. There is also an over-arching obsession for an experiment with the figure-ground. For Durbach Block, city and nature become one endless landscape to be redefined, reflected and remoulded. There is never an object in total isolation. Always in their architecture is relativity to further objects within a greater field. The goal is to continue the remaking and reflecting, the painting and the carving of that landscape – before the sea washes, or the wind wears those living walls away.

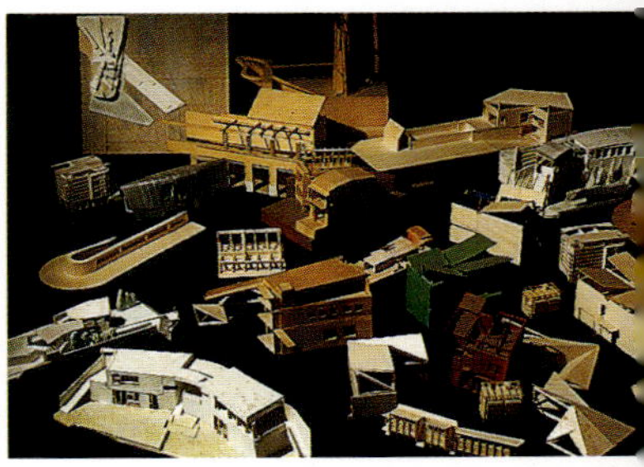

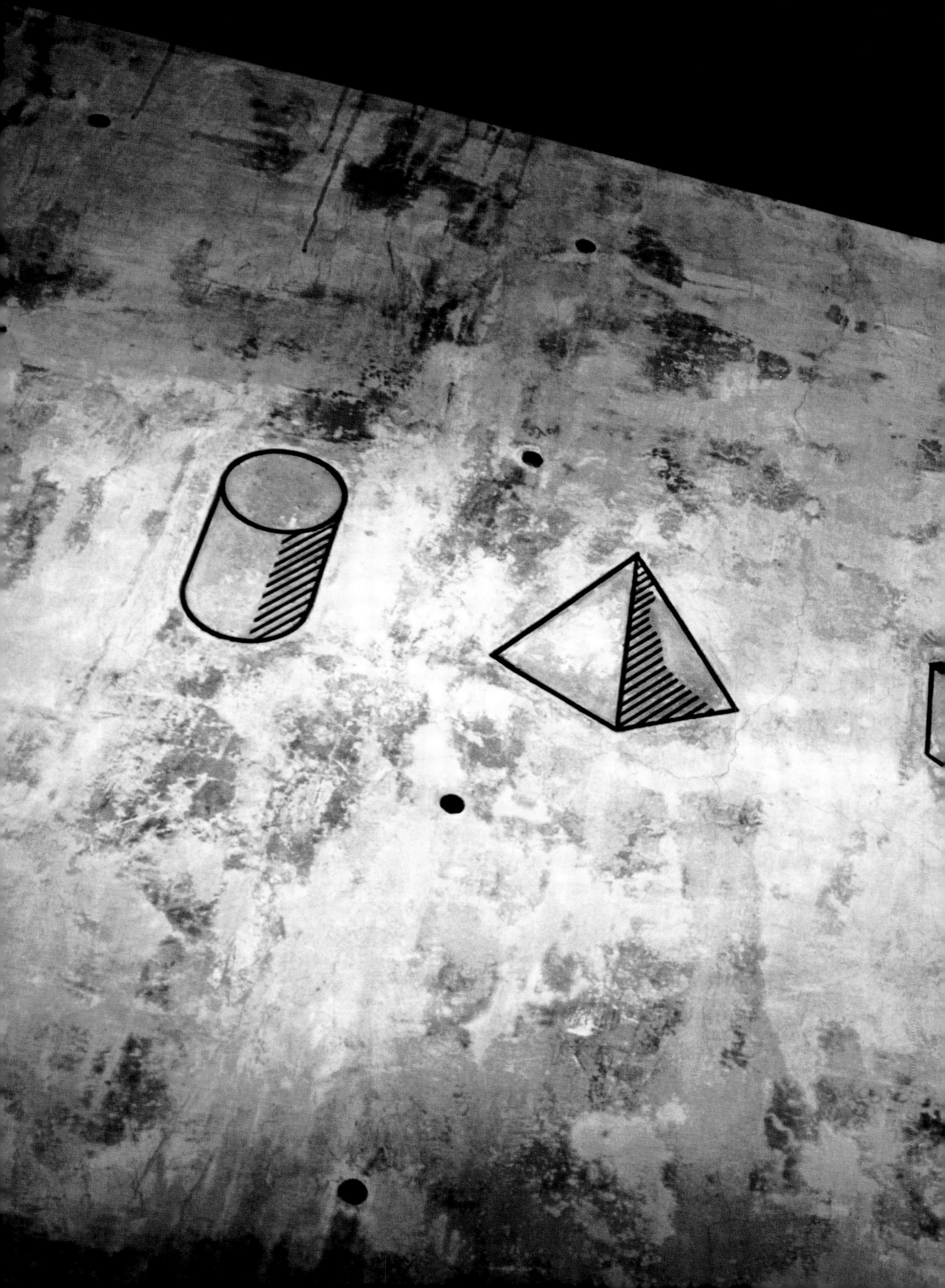

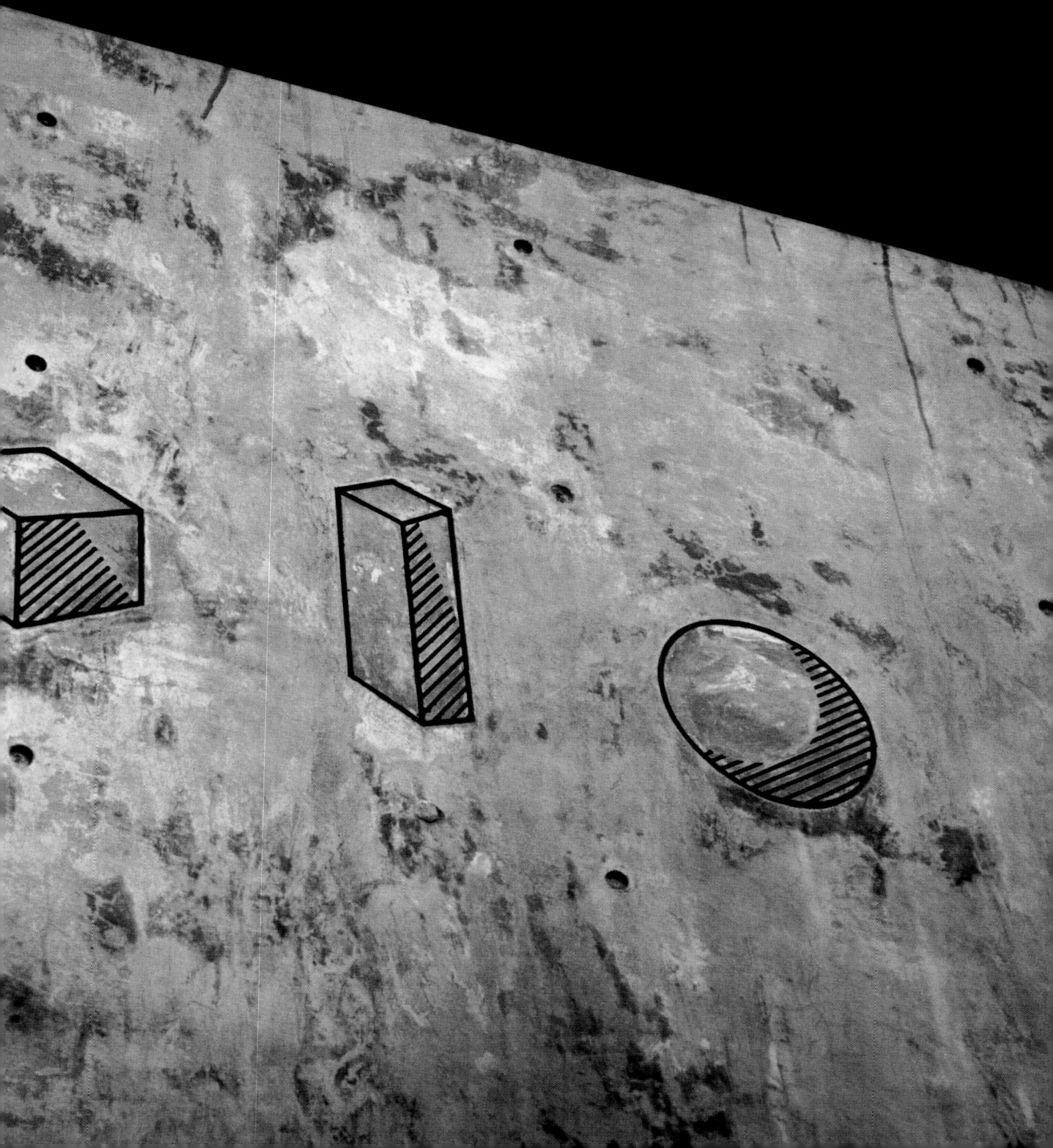

EXTENSIONS TO TUSCULUM FOR THE ROYAL AUSTRALIAN INSTITUTE OF ARCHITECTS

POTTS POINT, SYDNEY

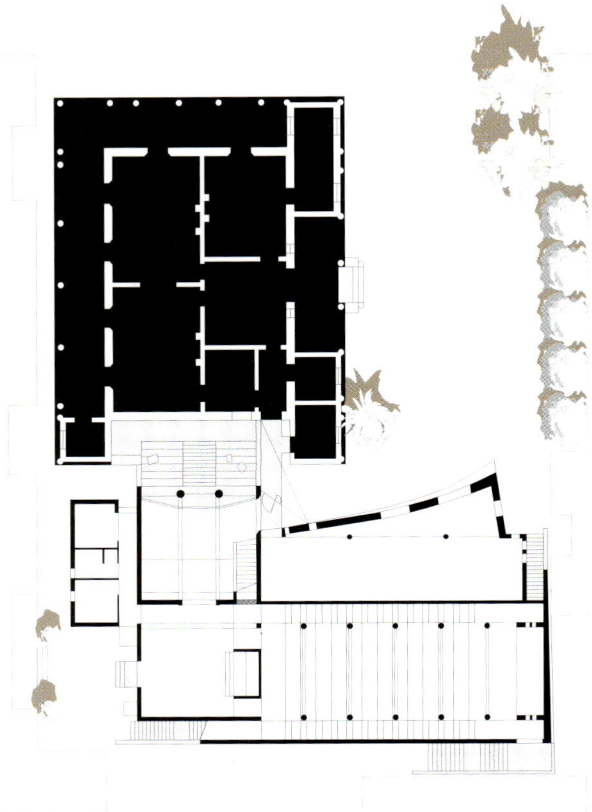

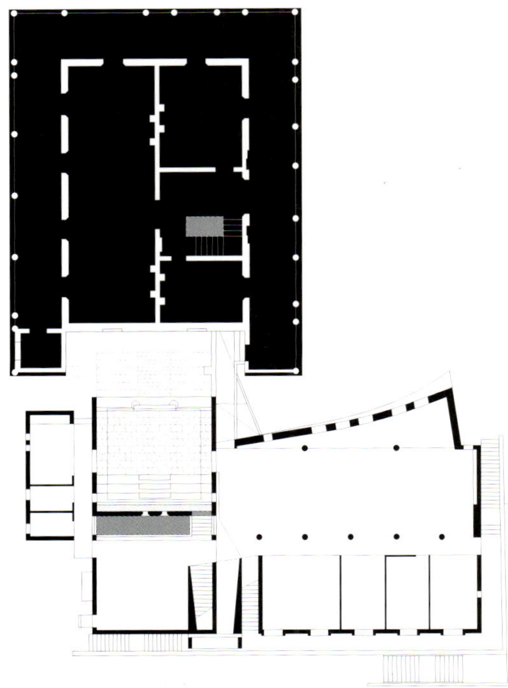

PROJECT TEAM Levine + Durbach; Neil Durbach, Harold Levine, Harry Margalit, Documentation in association with Allen Jack and Cottier

Date of completion 1988

This addition to the nineteenth century villa 'Tusculum', designed by John Verge, was won in a national architectural competition. The brief, for the Royal Australian Institute of Architects, called for a new wing housing office space, an auditorium and a gallery. The project raised two essential problems – how to add to an existing heritage villa; and how to accommodate a large and varied program on a small site. Implicit was an additional question: in a building for an institute of architects, is it possible to symbolise and summarise architecture itself?

The inclination to illustrate the idea of architecture in an institute building was resisted as being needlessly didactic. The formal possibilities of architecture itself seemed to diminish when required to serve as messenger or headline or storyboard of images. An empty court, a deferred entry sequence, a mute curved mulberry coloured wall, and an inversion of the villa's 'u' plan from perimeter to centre contribute to the reading of the villa as it was intended- a building in the round. Internally, the new wing is distinguished by a precise structural order, woven through an ensemble of spatial types.

FIRST FLOOR COURTYARD

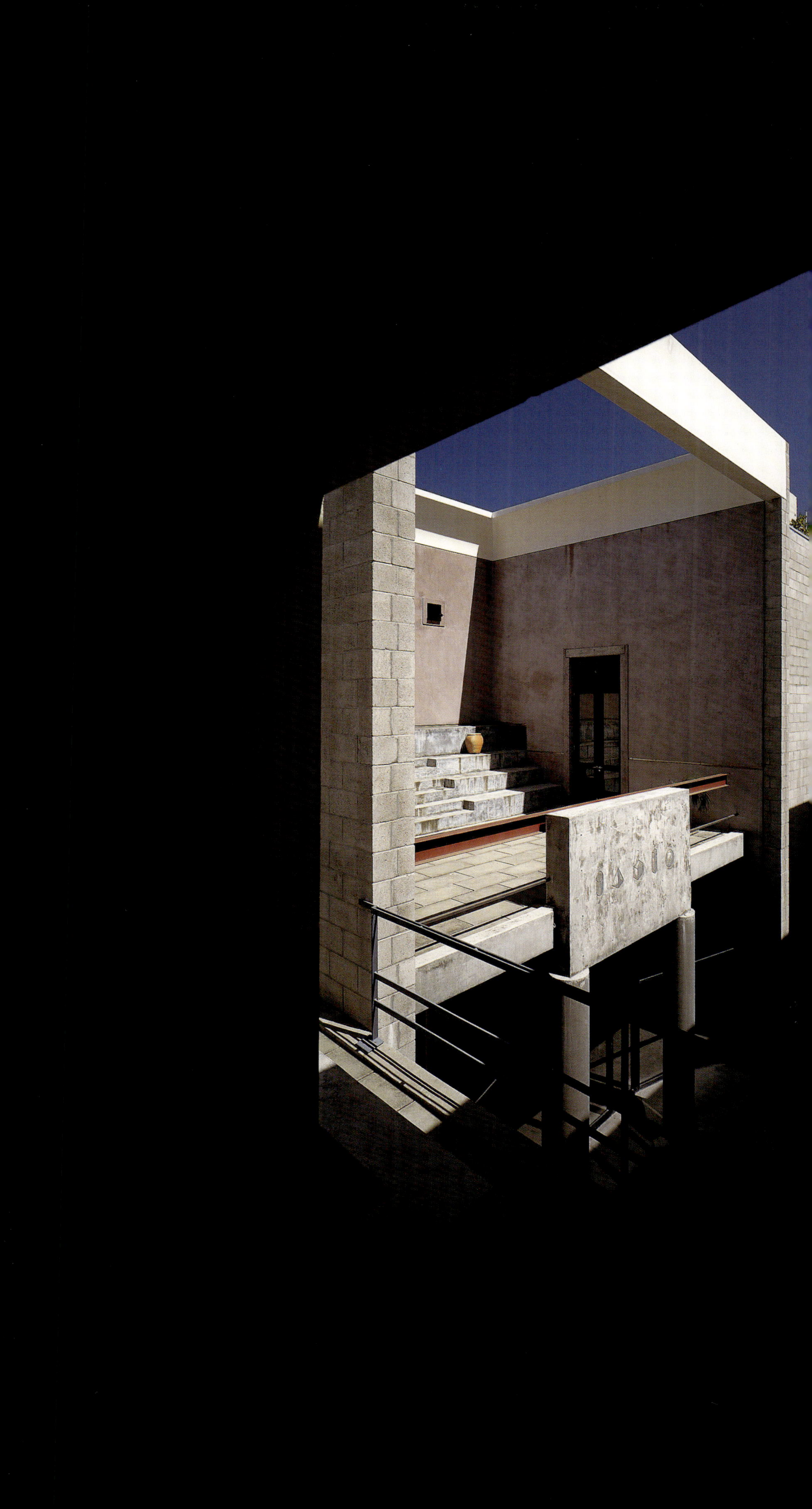

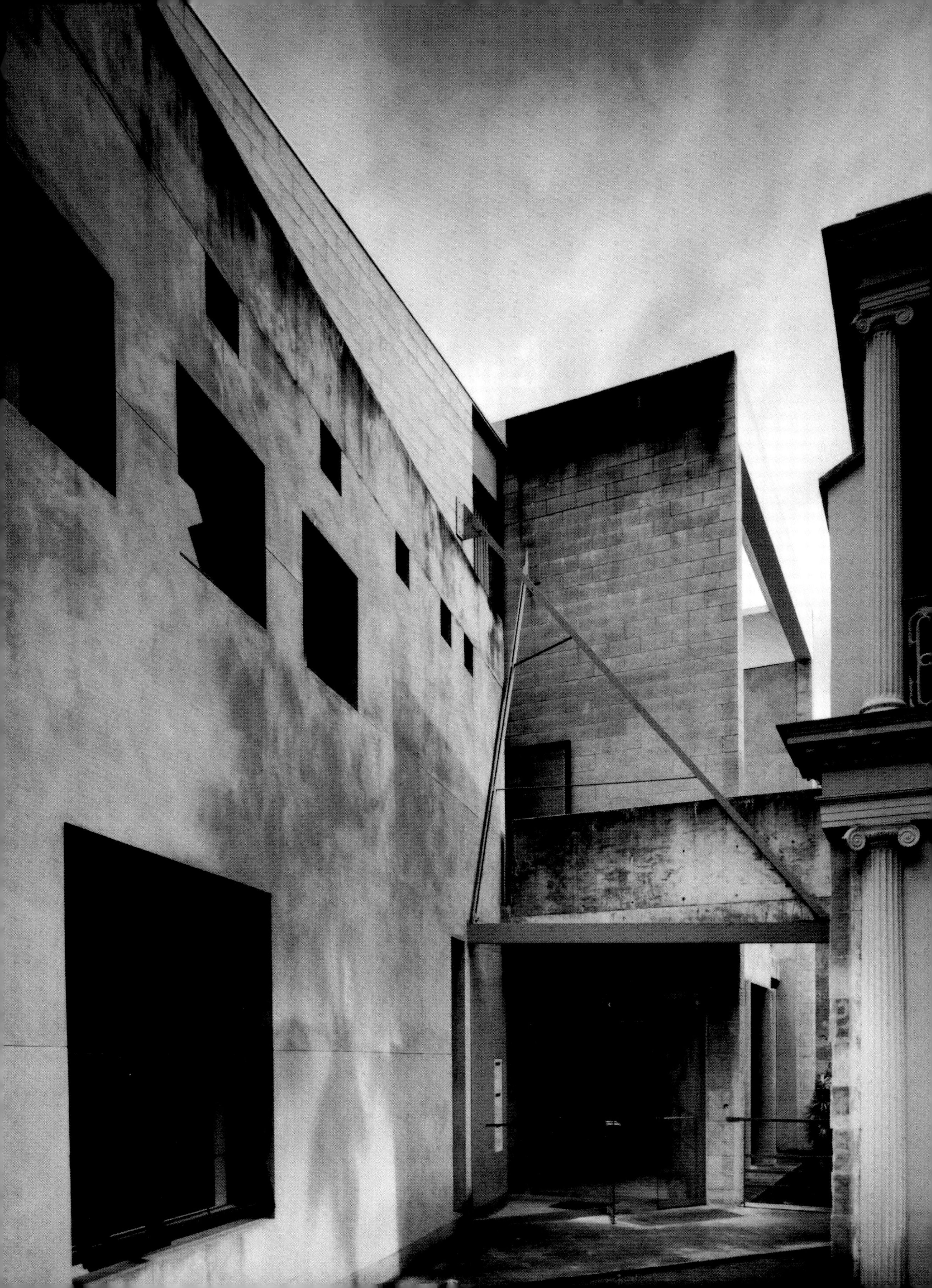

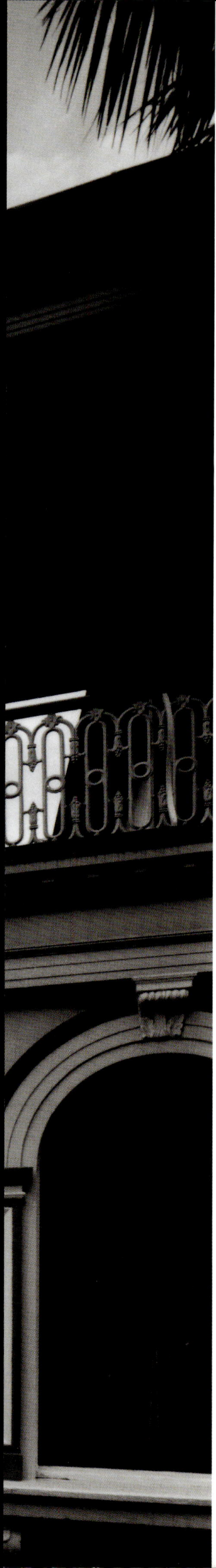

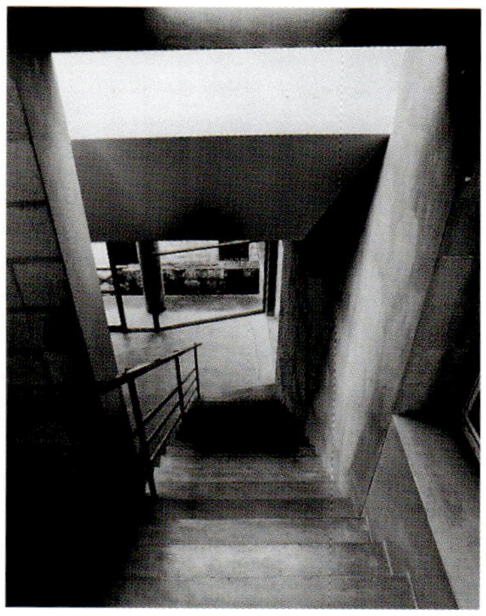

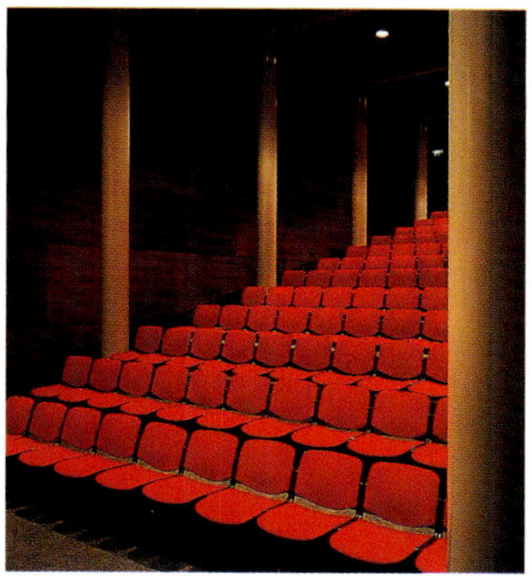

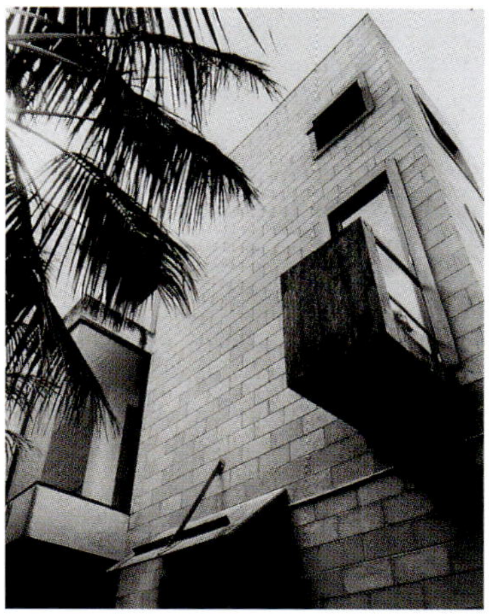

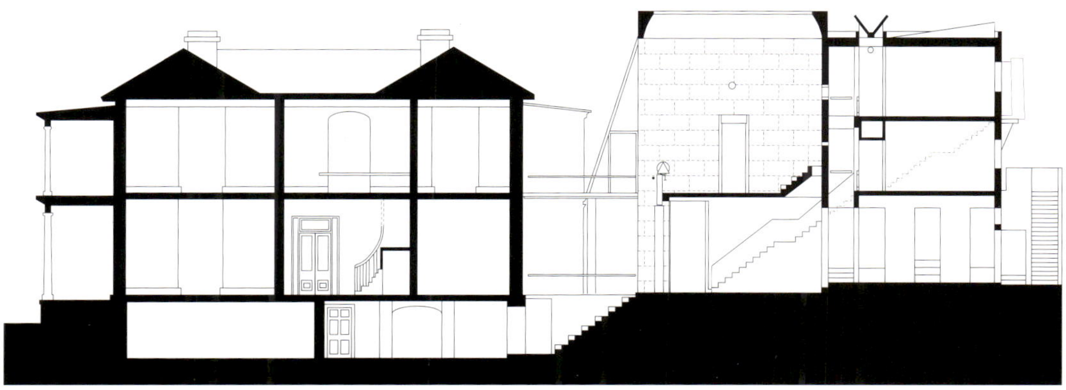

LEFT: NEW BUILDING WITH EXISTING VILLA TOP: STAIR FROM FIRST FLOOR CENTRE: THEATRETTE BOTTOM: SOUTH-WEST CORNER

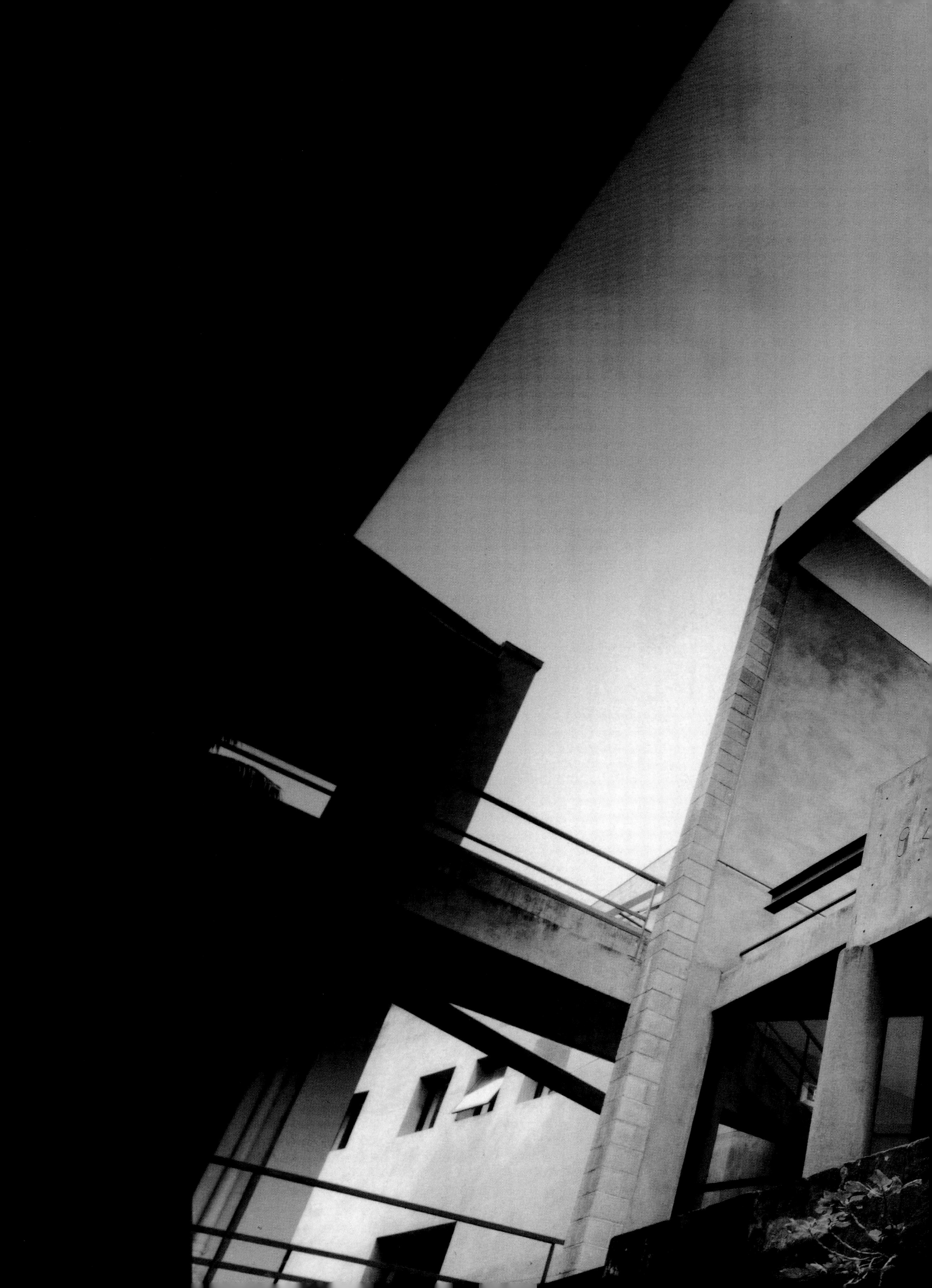

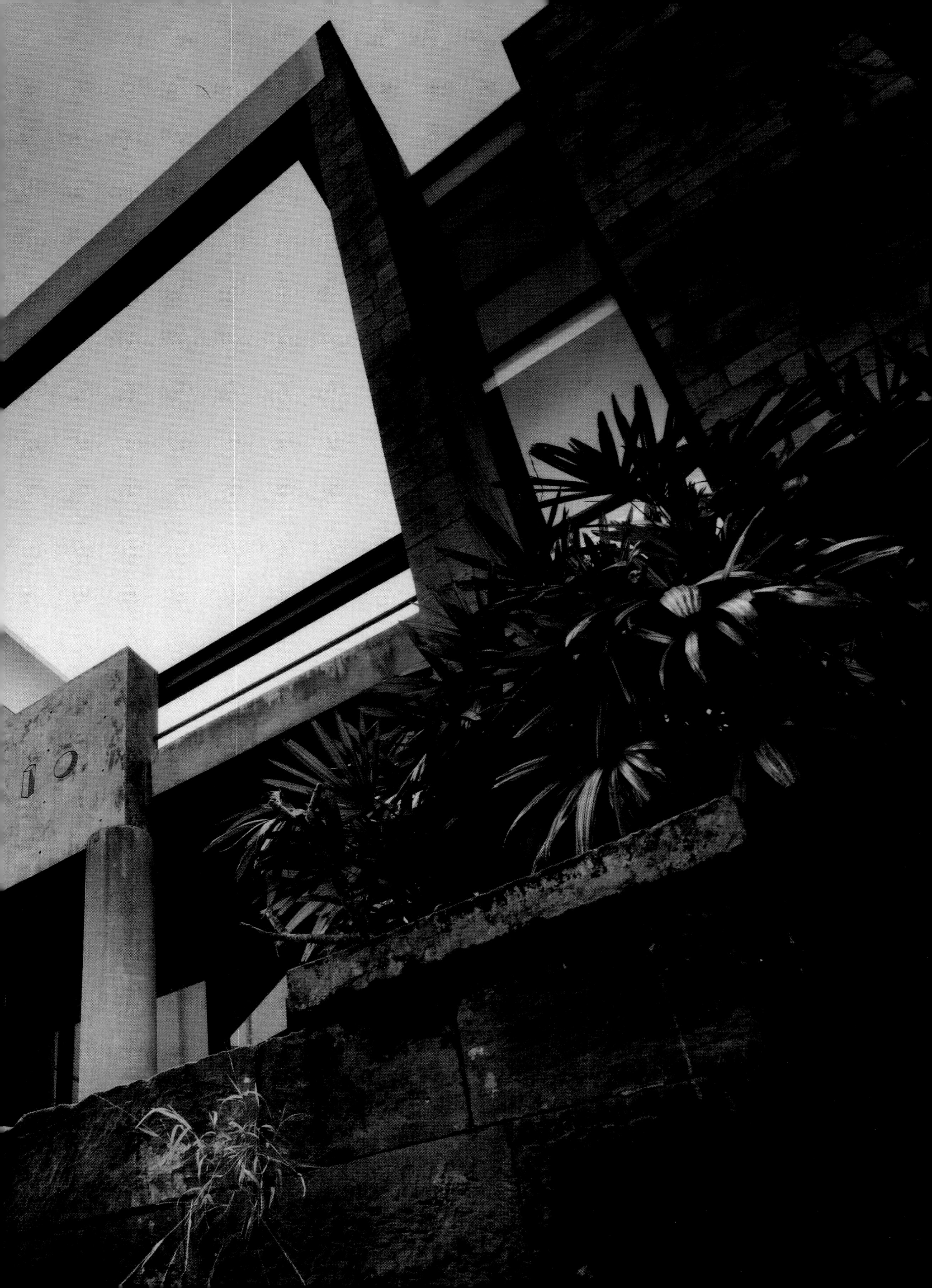
10

HOUSE VINH STRATHFIELD, SYDNEY

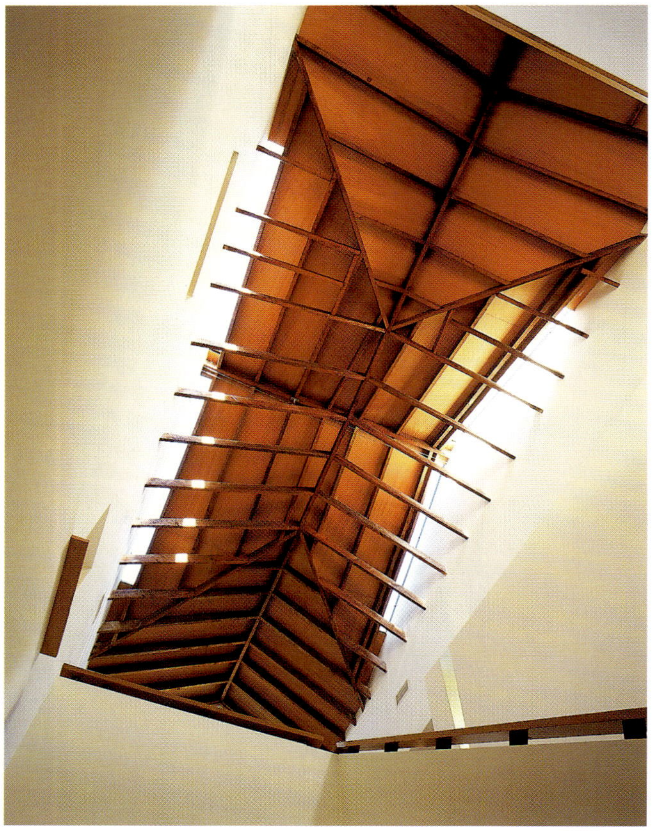

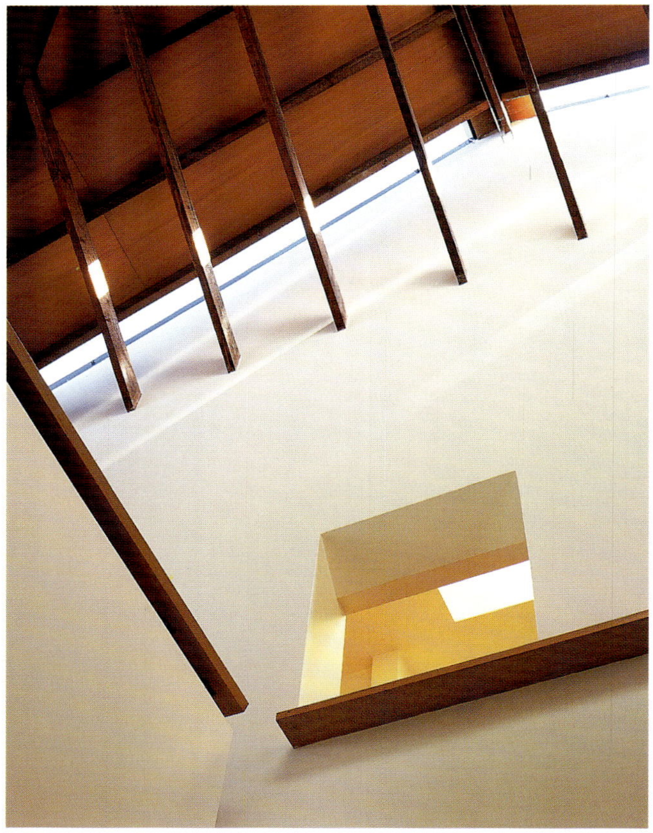

PROJECT TEAM Neil Durbach, Harold Straatveit

Date of completion 1990

Through an urban idea of distinction sparingly applied, an ordinary suburban house was transformed by the clearing of a single space to create a moment of light in the middle of density.

A reconfigured staircase relinked the existing rooms at all levels. Secondary framing was added to the exposed existing roof construction, crafting a complex and ambiguous structure. Inspired by the indeterminate space described between the two chords of the Harbour Bridge, the roof deliberately flouts the modernist credo of truth to structure. Strips of light are filtered through this framework, tracking the sun's passage across the sky. Set plaster, white glass and timber form a smooth background to the supple vigour of the ceiling.

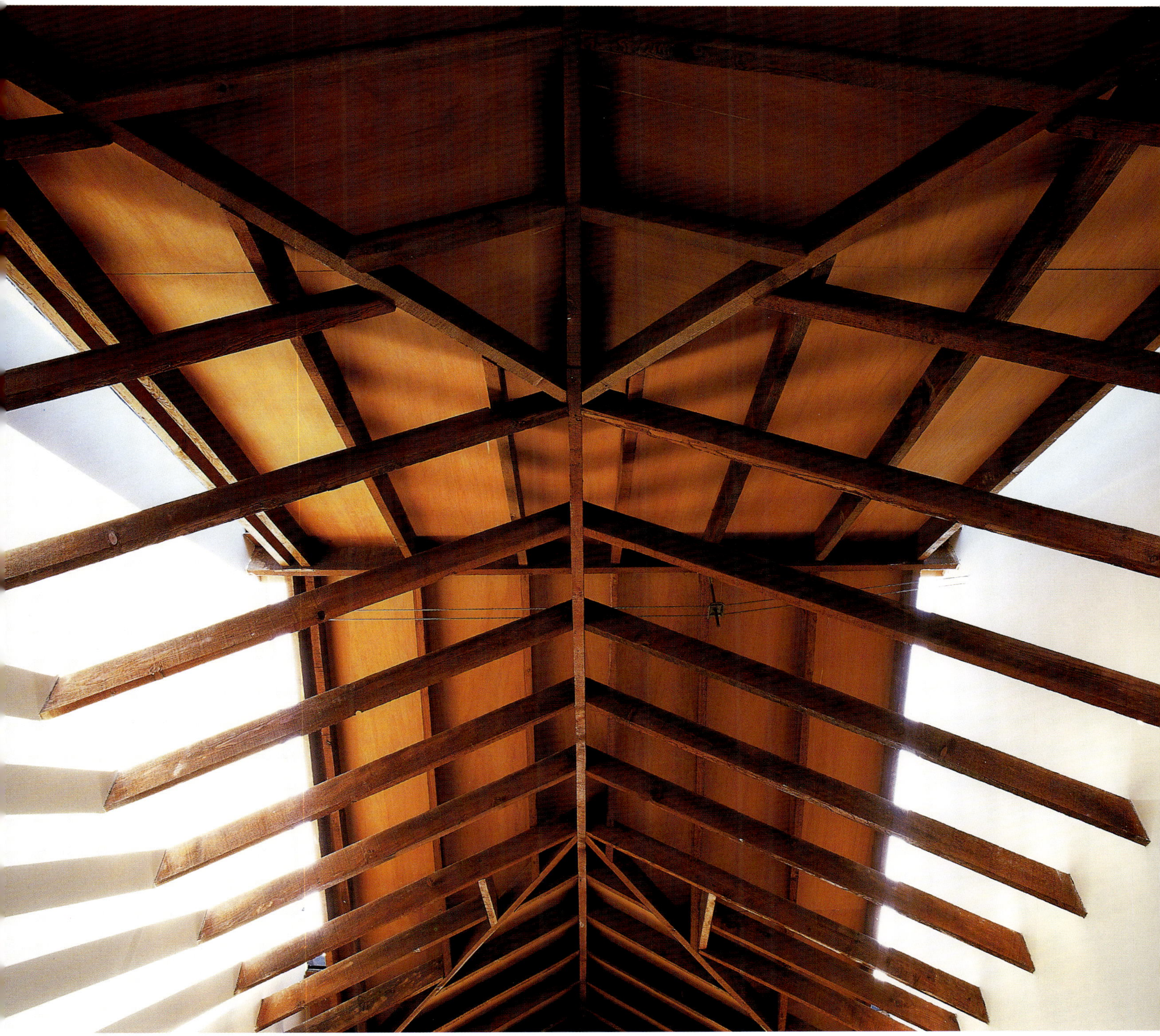

OPPOSITE LEFT: CEILING FROM ENTRANCE CENTRE: GALLERY ABOVE: CEILING FROM GALLERY

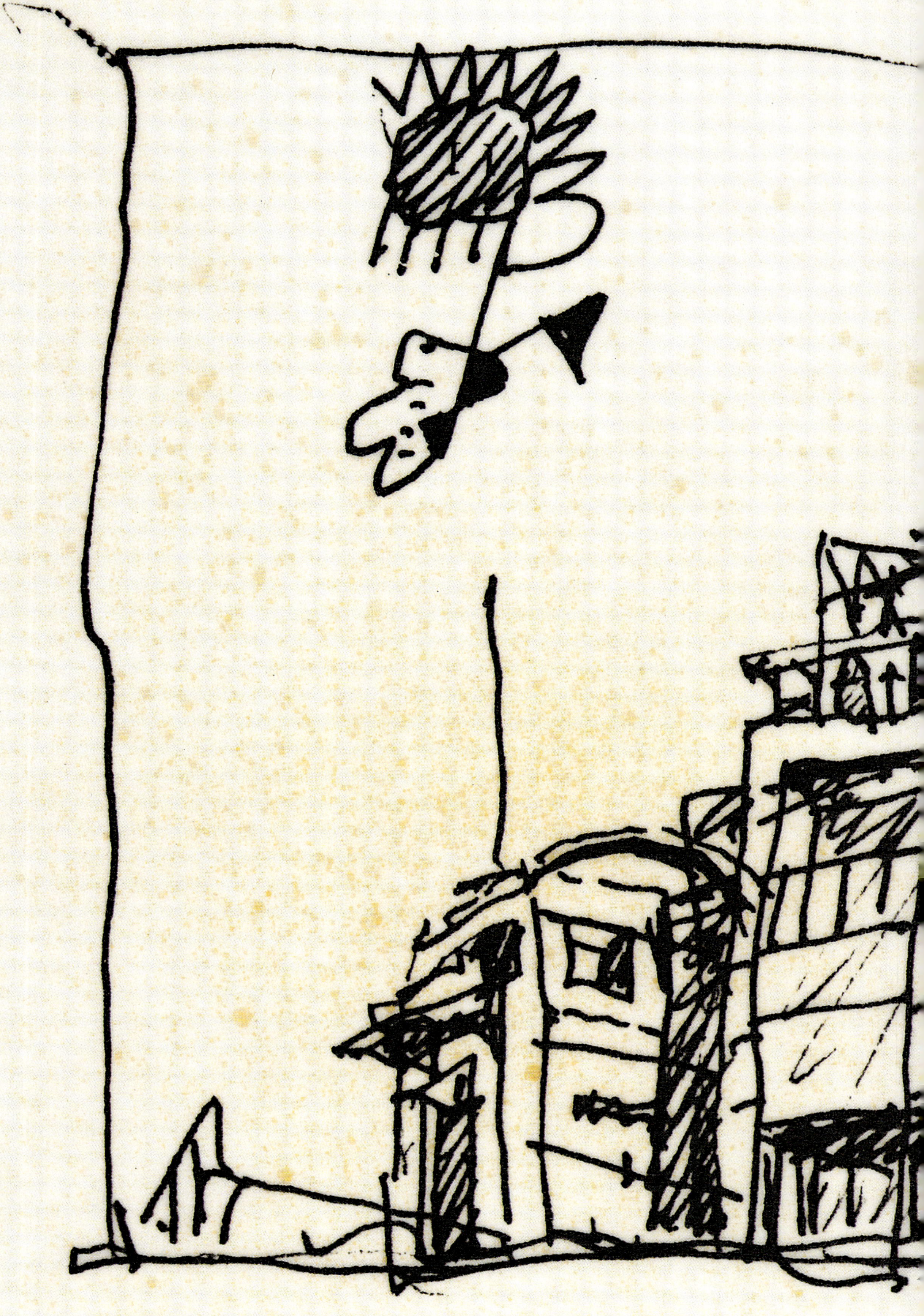

PROJECT THREE

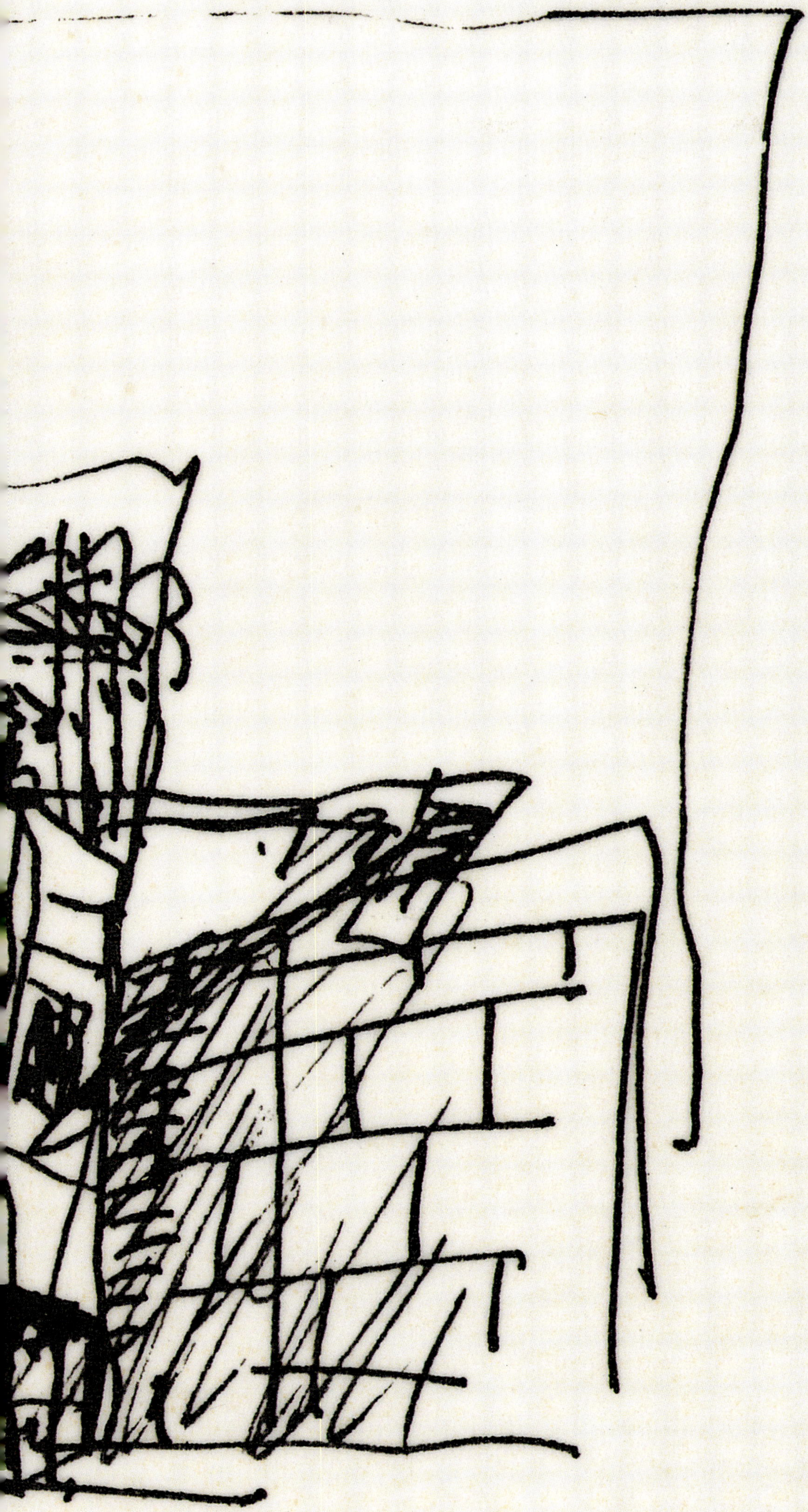

DARLING POINT HOUSE DARLING POINT, SYDNEY

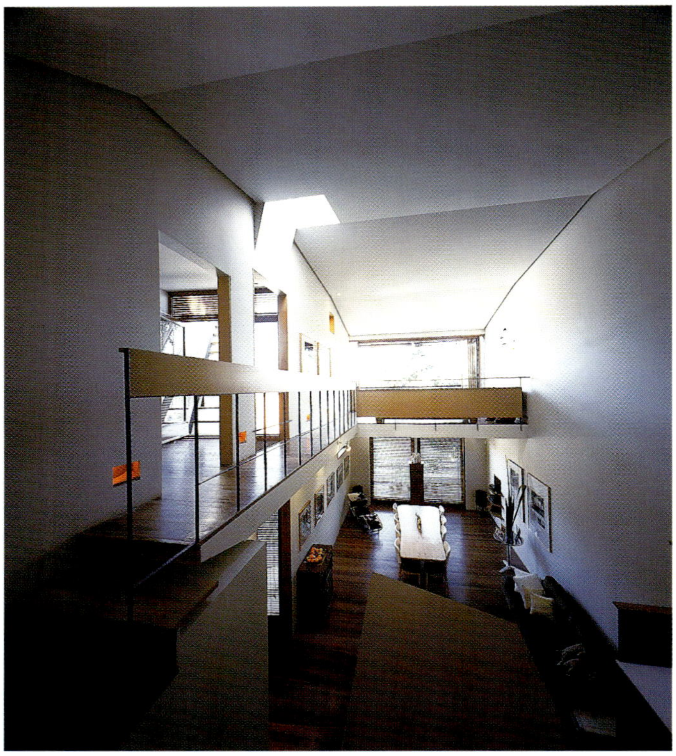

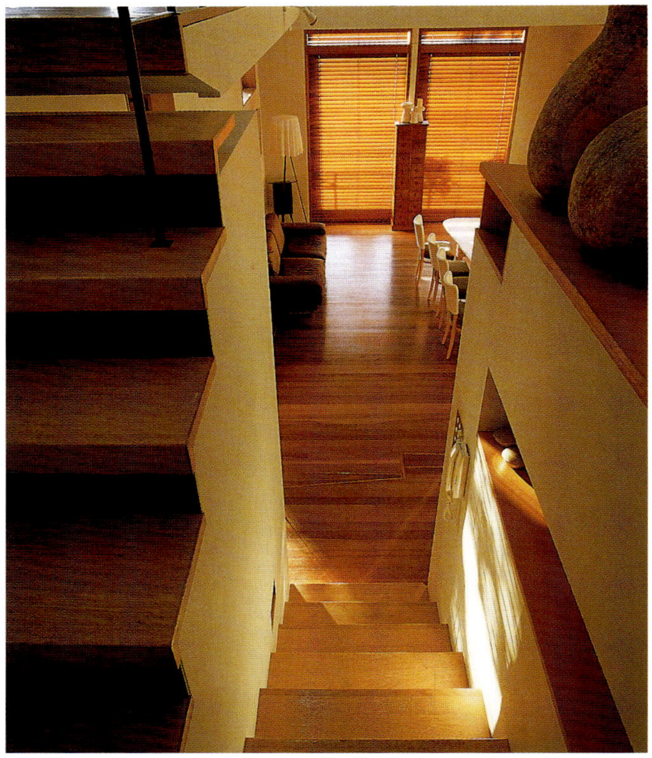

A tiny wedge-shaped site, bounded by a very old stone retaining wall. The site is secretive, connected to the street by a narrow staircase. Arriving, it is surprising to find a building there, like a ship in a bottle – a large and delicate object passed through an improbably small opening.

Although the two angled pavilions house a simple program – two bedrooms, two bathrooms and two living rooms, the house is organised by a series of complex external spaces at different levels. The entry court flares to hold late afternoon sun in winter. The kitchen court is shaded for summer, with a natural stream captured in a sandstone pond. A walled court harbours a slender lap pool with a triangular timber deck above. A roof terrace high over Rushcutters Bay is dominated by a gigantic pink bougainvillea, carefully tended during construction and trained onto an angled pergola.

These spaces are linked by a series of external elements; a dangerously steep staircase, a stick like bridge, and a series of stone ledges, woven through the interior of the house. The house becomes a description of that place, exaggerating the experience of a tiny house into an unfolding and endless promenade through and across the inside and outside spaces, imbued with the pleasure of use.

PROJECT TEAM Neil Durbach, Andrew Cortese, Camilla Block, Robin McInnes Date of completion 1992

LEFT: LIVING ROOM AND GALLERY FROM EAST CENTRE: STAIR AND LIVING ROOM RIGHT: LIVING ROOM FROM GALLERY

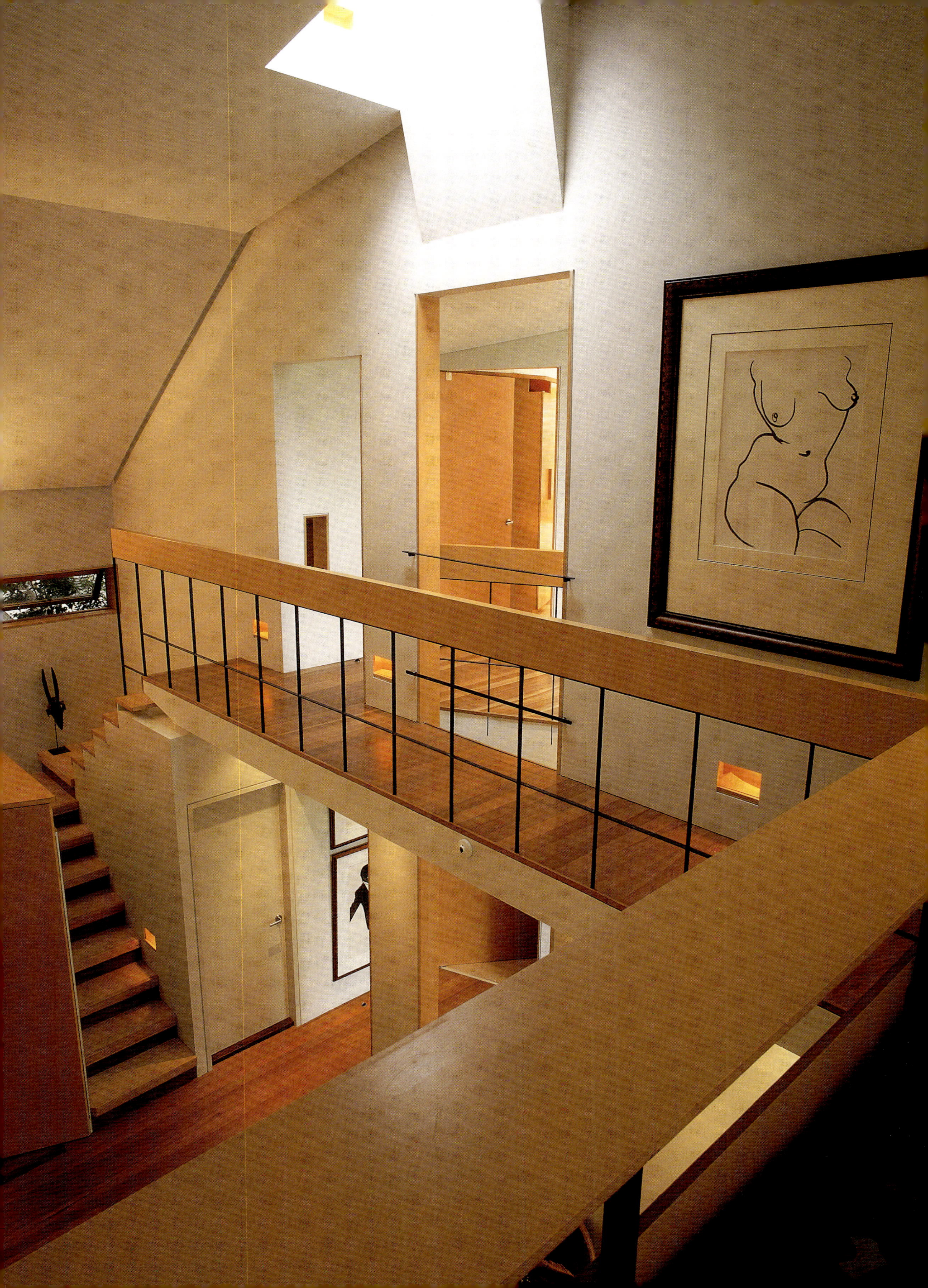

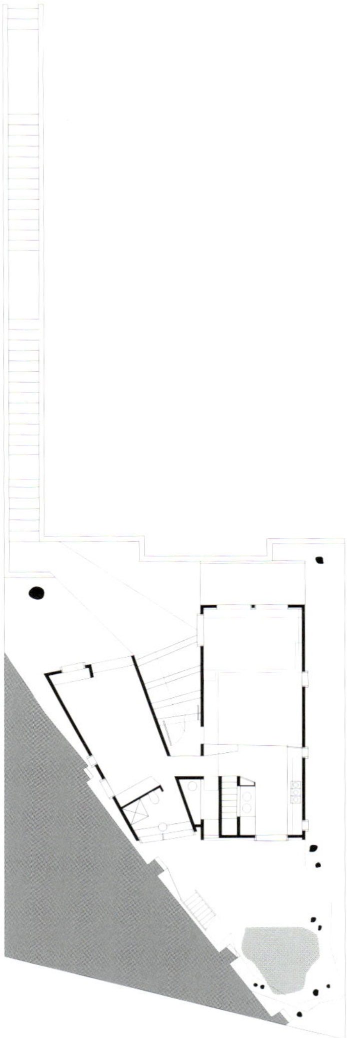

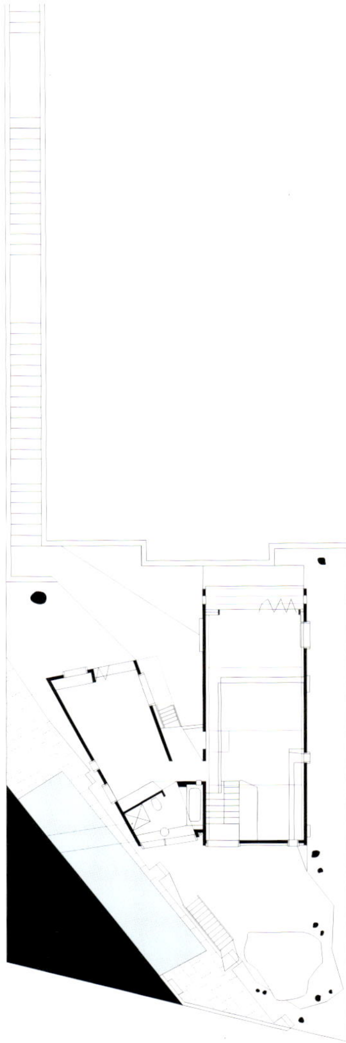

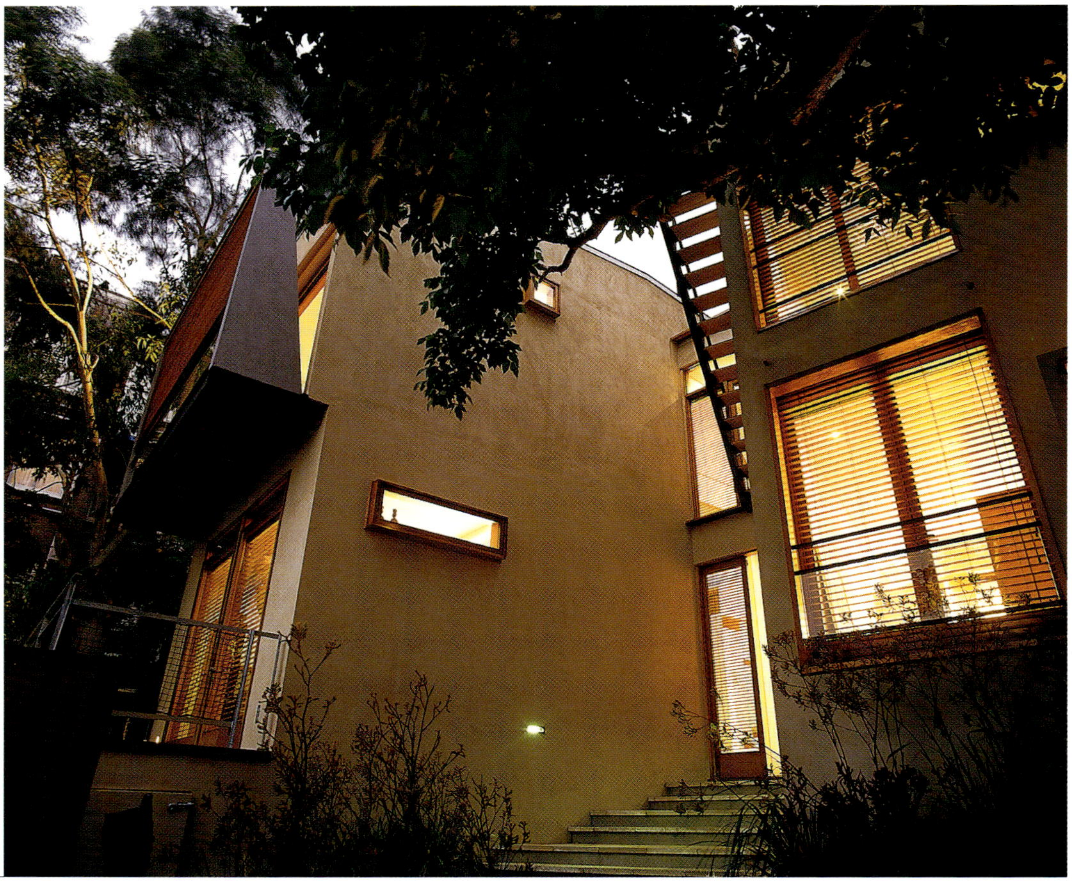

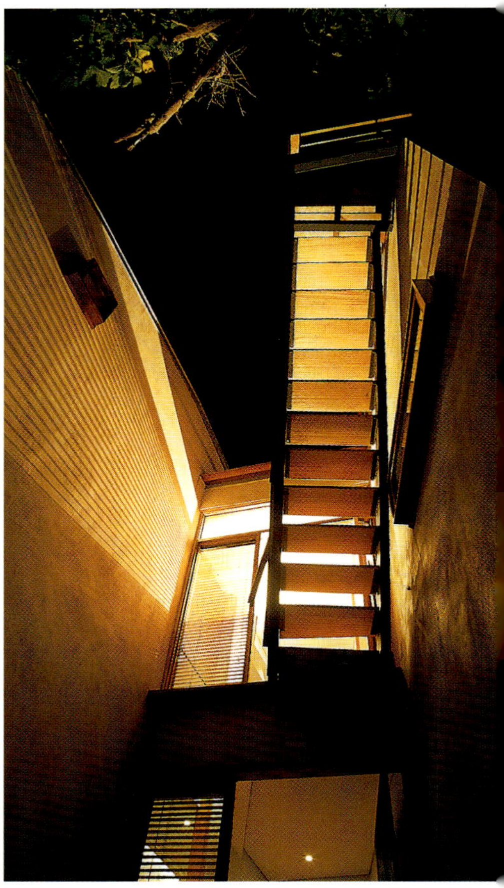

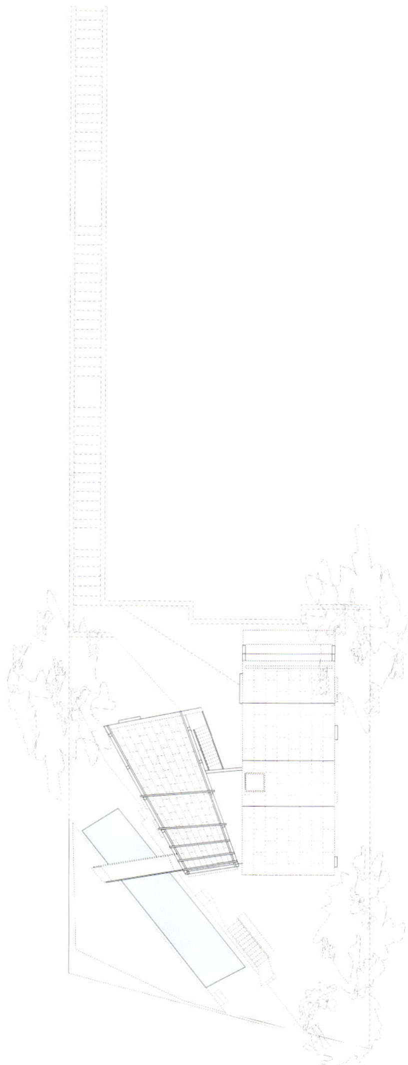

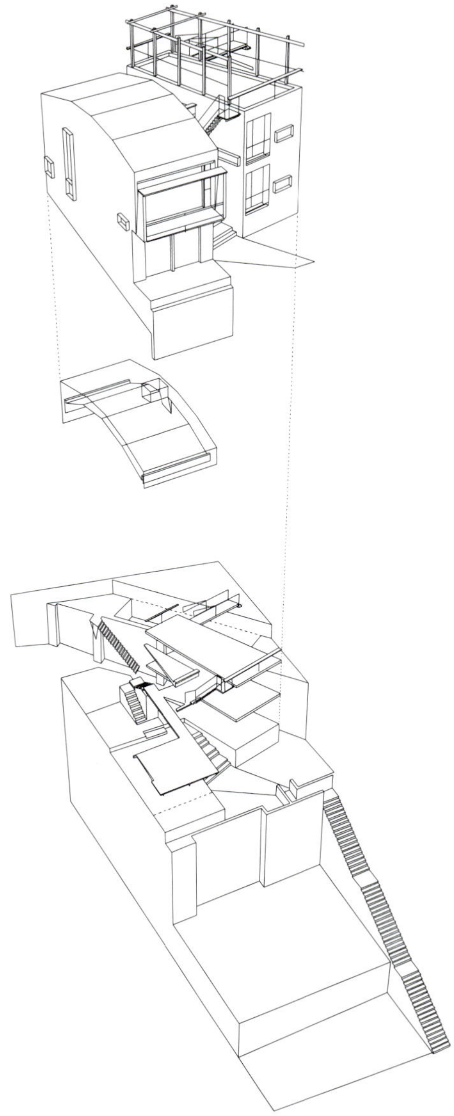

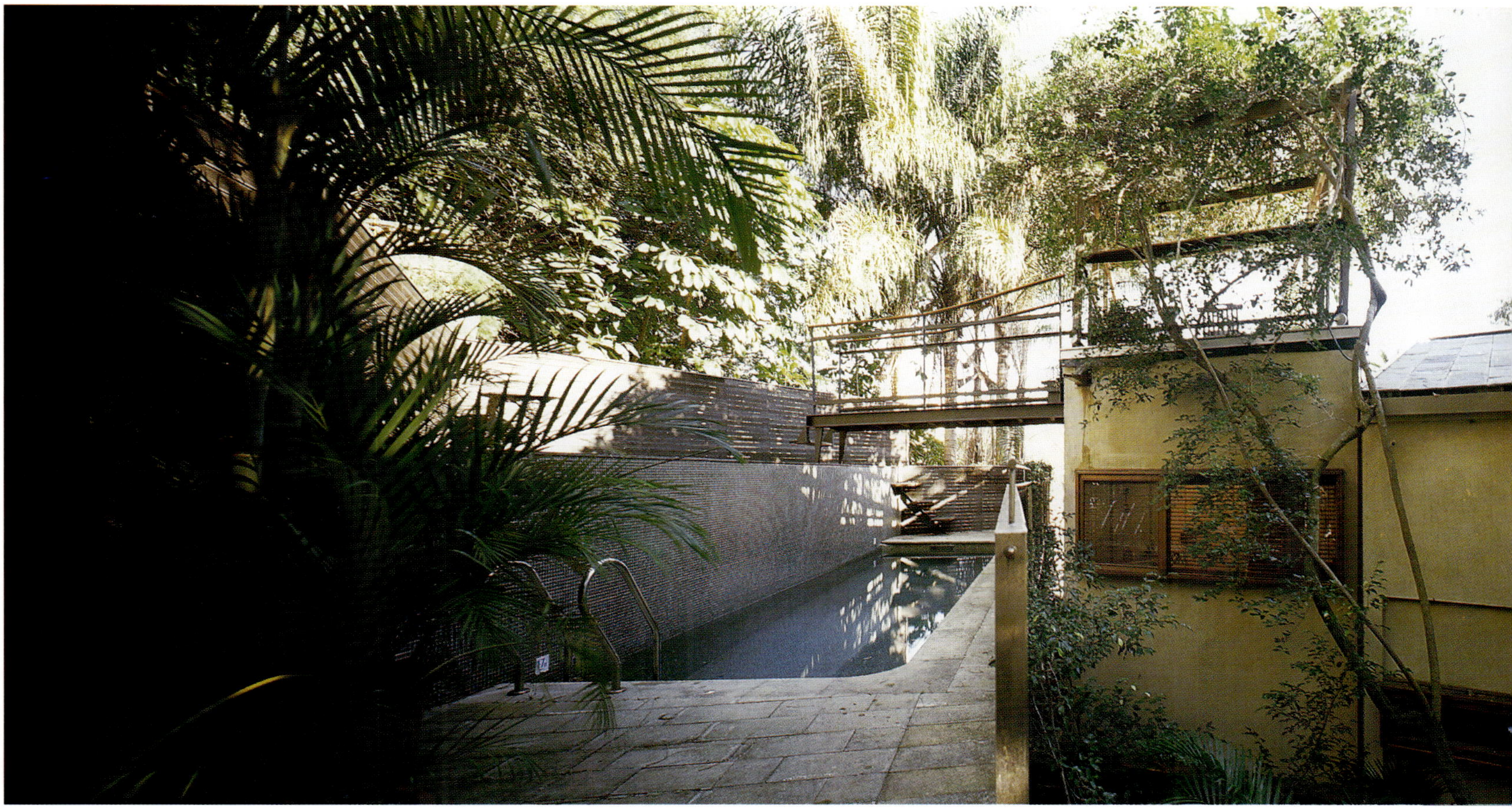

LEFT: ENTRANCE FROM SOUTH WEST CENTRE: FIRST FLOOR EXTERNAL STAIR ABOVE: ROOFTOP LAP POOL

MODEL URBAN HOUSING ST. CLAIR

A pocket of density inserted into the mat of suburbia. The scheme retains the ideals of the suburb: defined houses, private gardens and individual vehicles. Higher density is achieved through a communality of spaces and amenities. The insertion of garden courts as an intermediate spatial type leavens the density, creating small commons for access, play and rainwater dispersal.

PROJECT TEAM Neil Durbach, Camilla Block, Nick Murcutt, Andrew Cortese, Harry Margalit 1993

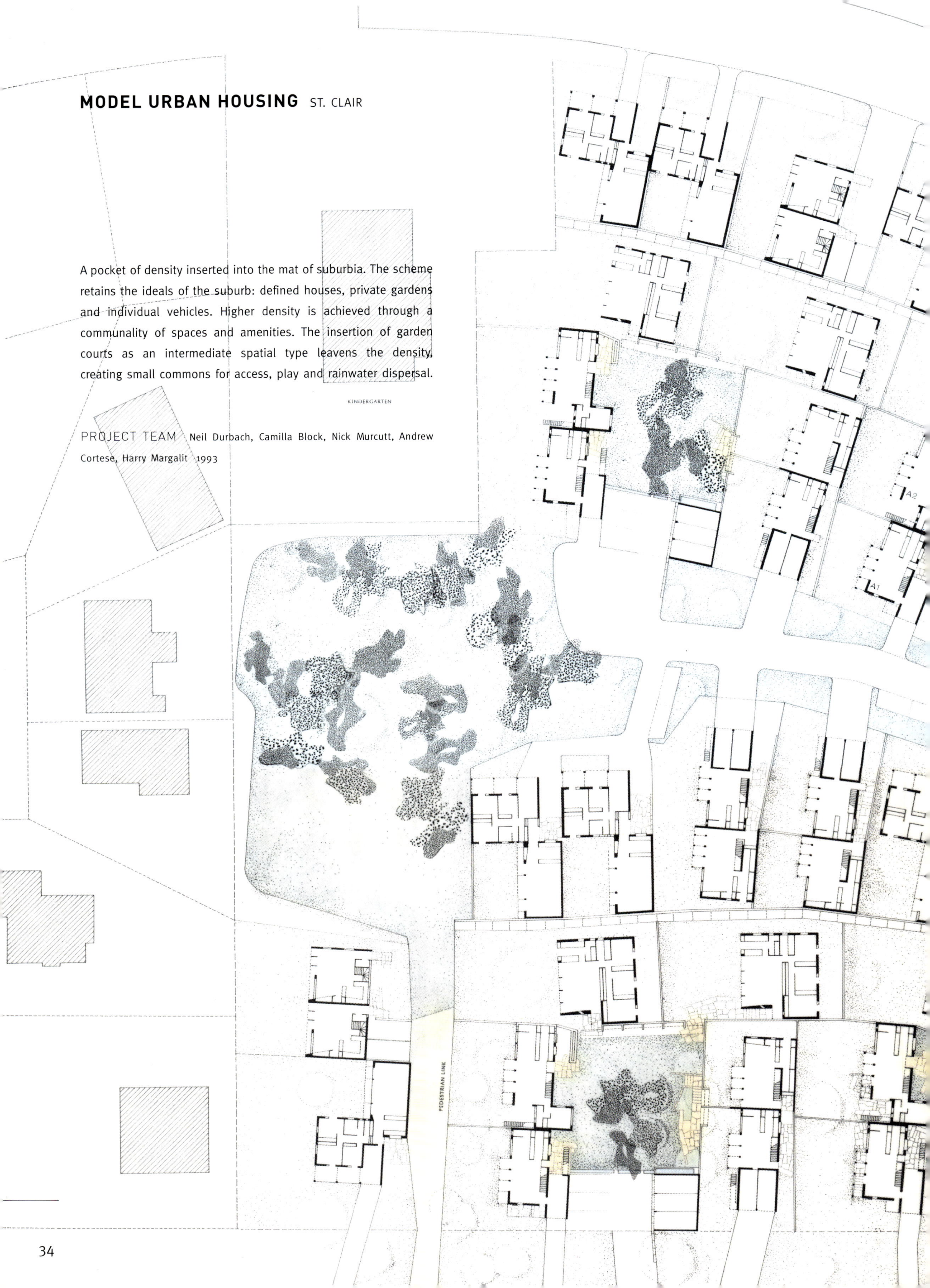

PROJECT FOUR

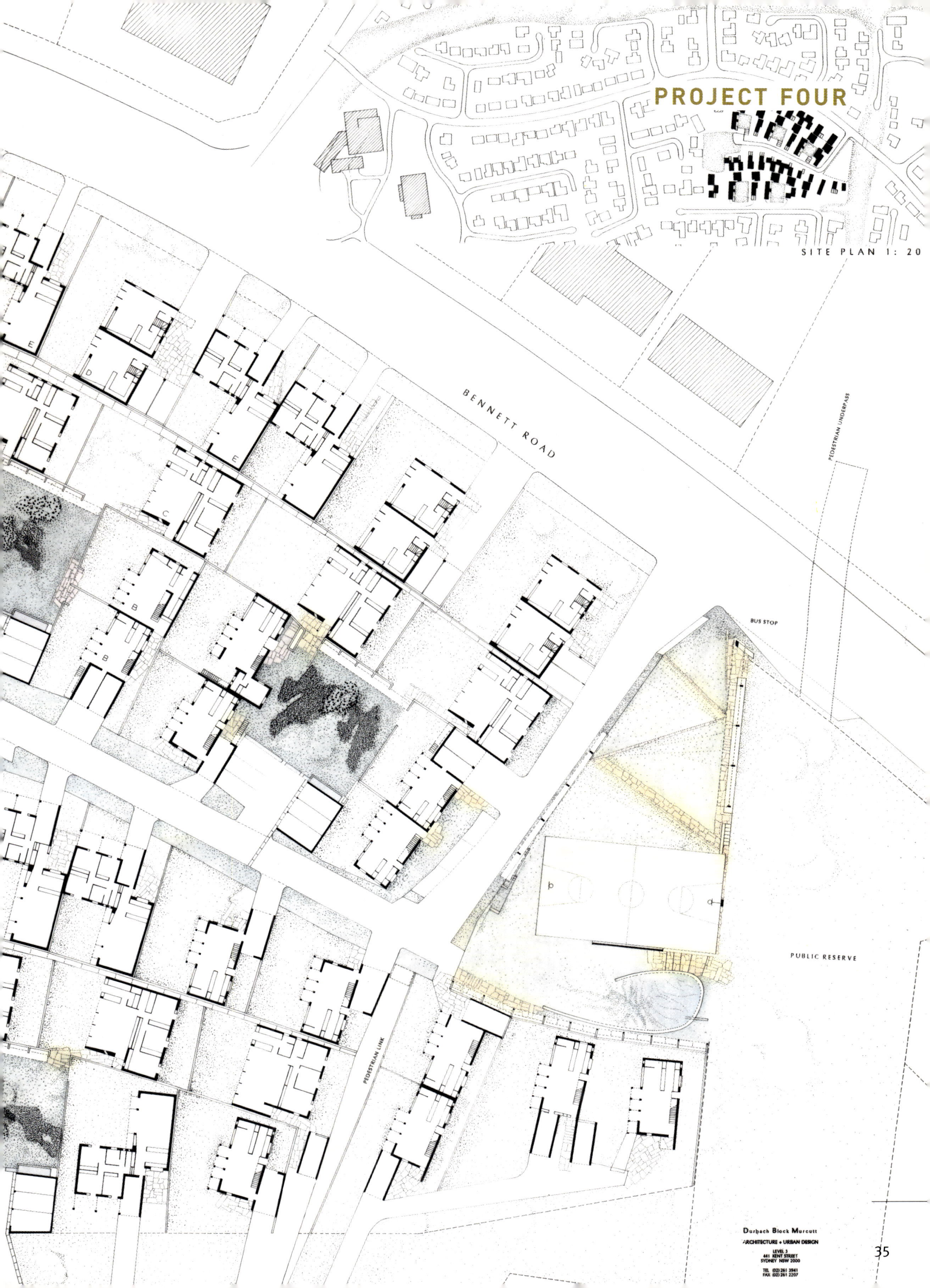

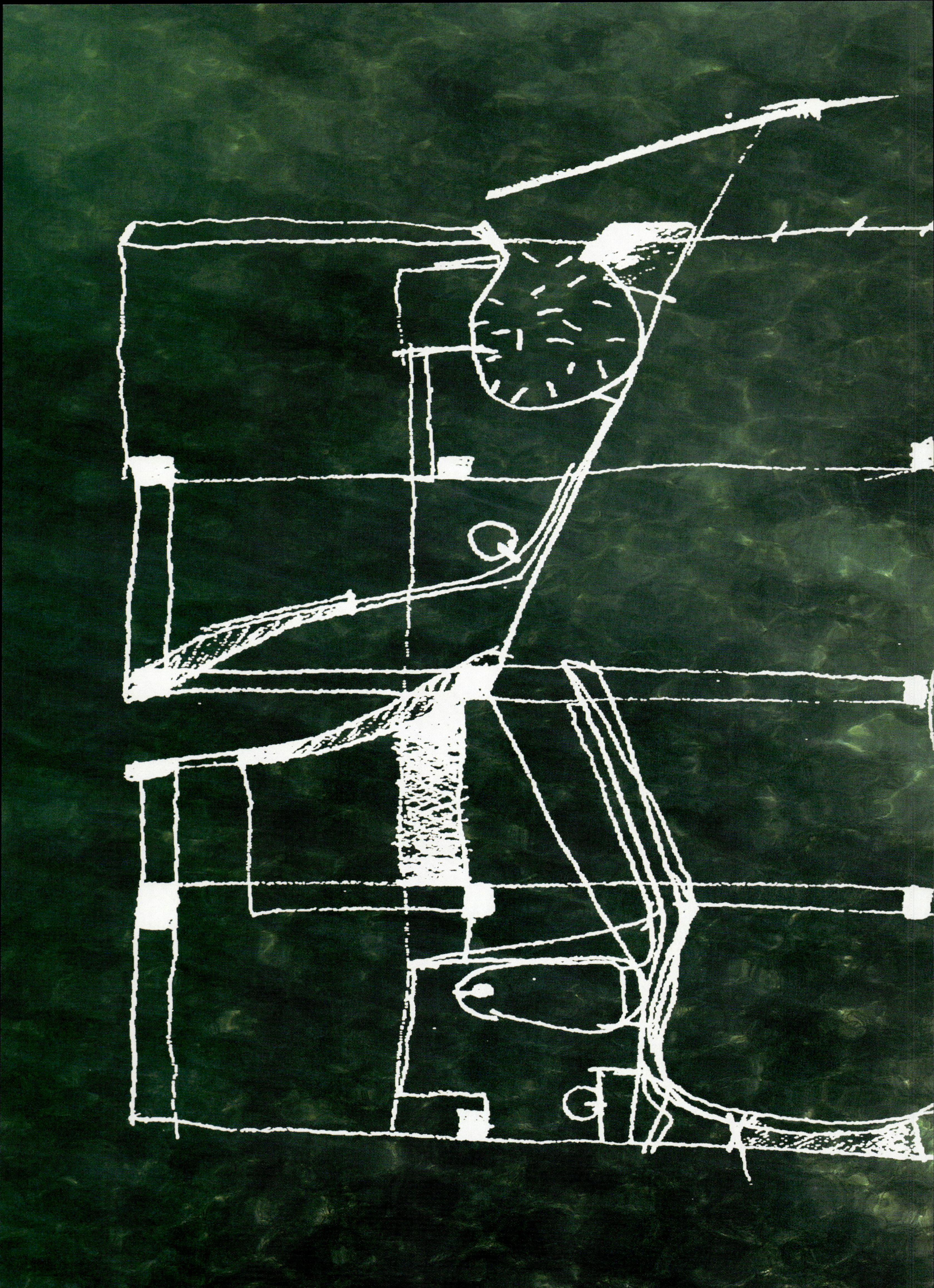

PROJECT FIVE

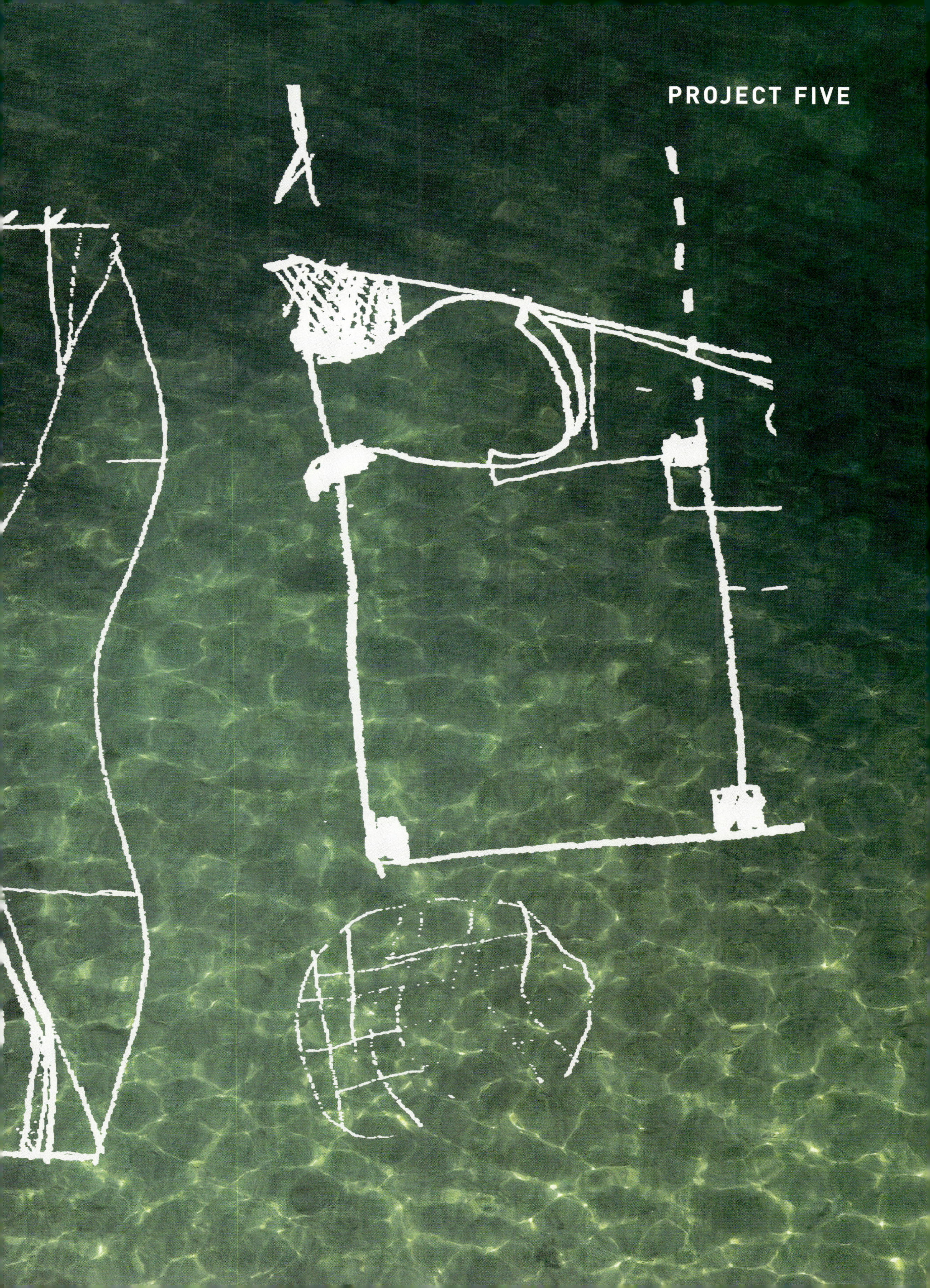

EINFELD APARTMENT DARLING POINT SYDNEY

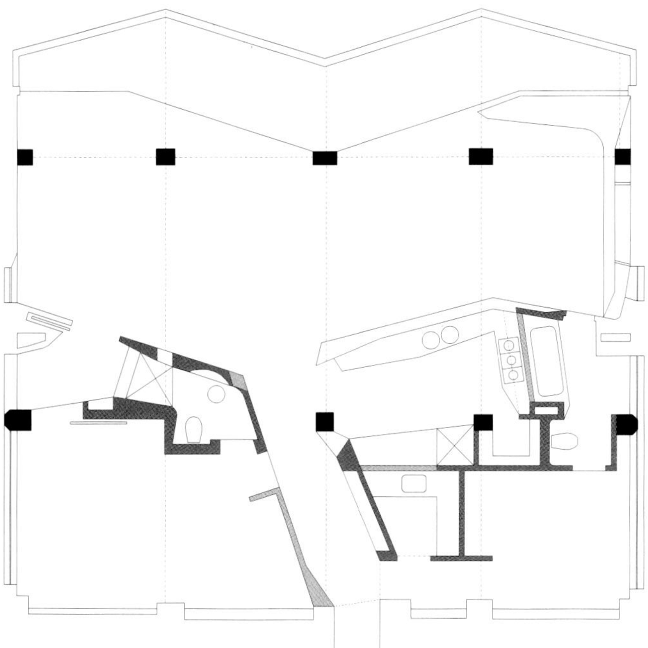

PROJECT TEAM Neil Durbach, Camilla Block, Lisa Le Van

Date of completion 1995

A seventh floor apartment in a 60s facebrick building, so close to the water's edge that would no longer be allowed. To the south lies the enclosed cove and park of Double Bay, green and cosy blue: to the north, an expanse of harbour and sky, bright and scale-less.

The existing apartment was a symmetrically cellular mess. A buried dining room, a remote kitchen and blinkered living room. The brief called for one living area large enough to have family dinners for fourteen, but generally for two people. A central kitchen, suitable for serious and pleasurable cooking.

With all non-loadbearing walls demolished, the site was simply two concrete plates, ceiling and floor; a fixed perimeter with angling glazed walls and a balcony to the east, unceremoniously abutting a rational structural grid of columns and beams. All living was brought forward to the balcony - one large smooth space, seamless with the water on a good day, precipice-like in dramatic weather. Geometries are loose, circumstantial, bending to accommodate the exigencies of the site: the grid, ducts, windows, and the glazing line.

The bedrooms face north or south, with floors raised for views at bed height. A bathroom retreats from the perimeter of the building to form an atrium, a calm suntrap when the weather blows south, additional sun for the living room, a spacious outside shower, and a secret circuit through the apartment.

The kitchen is coloured glass with iridescent laminate, liquid-like shades and intensities of green, green-yellow and green-blue. The lighted perimeter is drawn into the depth of the building, with light recycled from room to room.

The study joinery curves and contains space, emulating the smooth outline of the cove, while the geometry of the living and dining area broadens with the expanding view.

KITCHEN

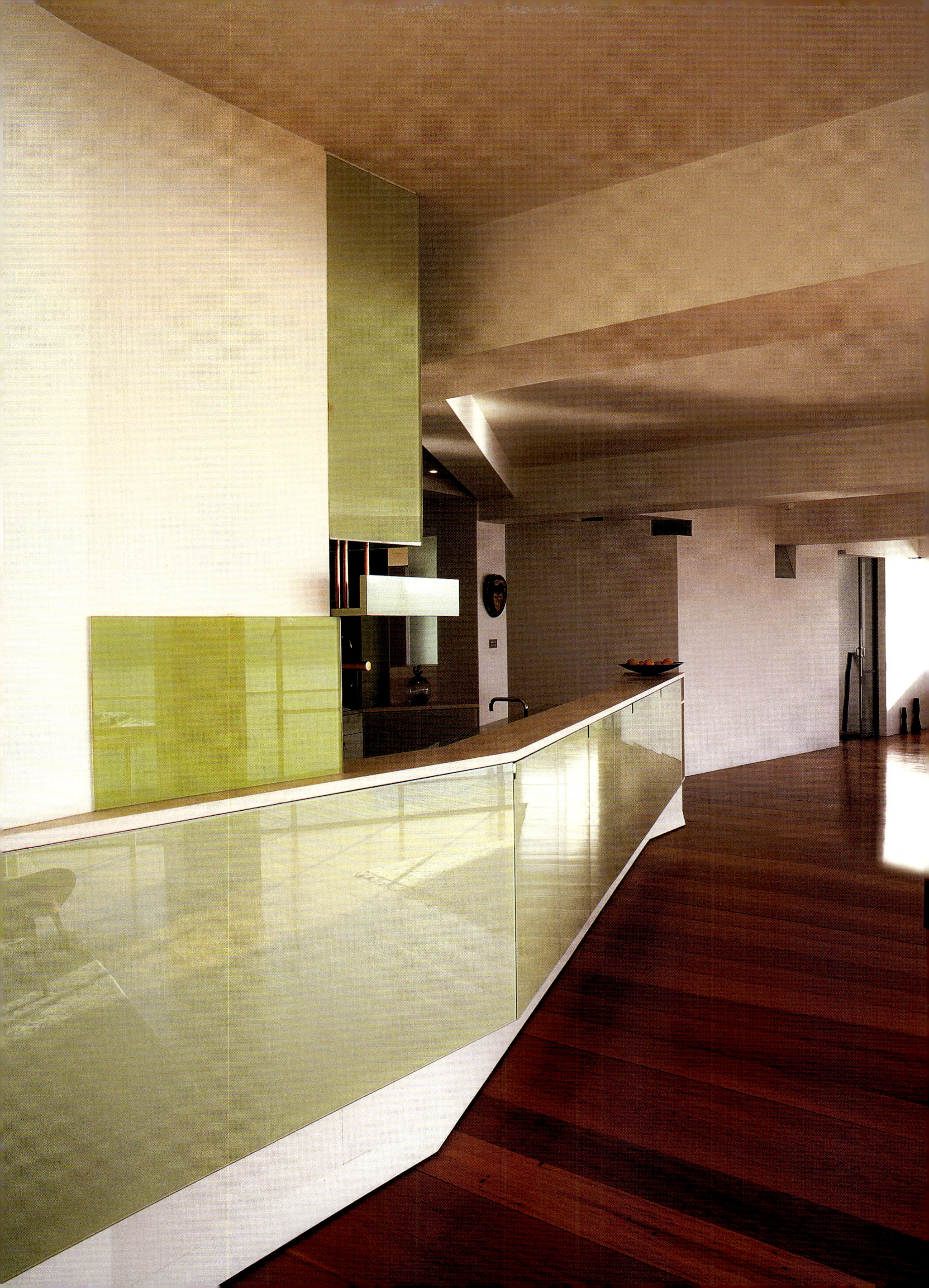

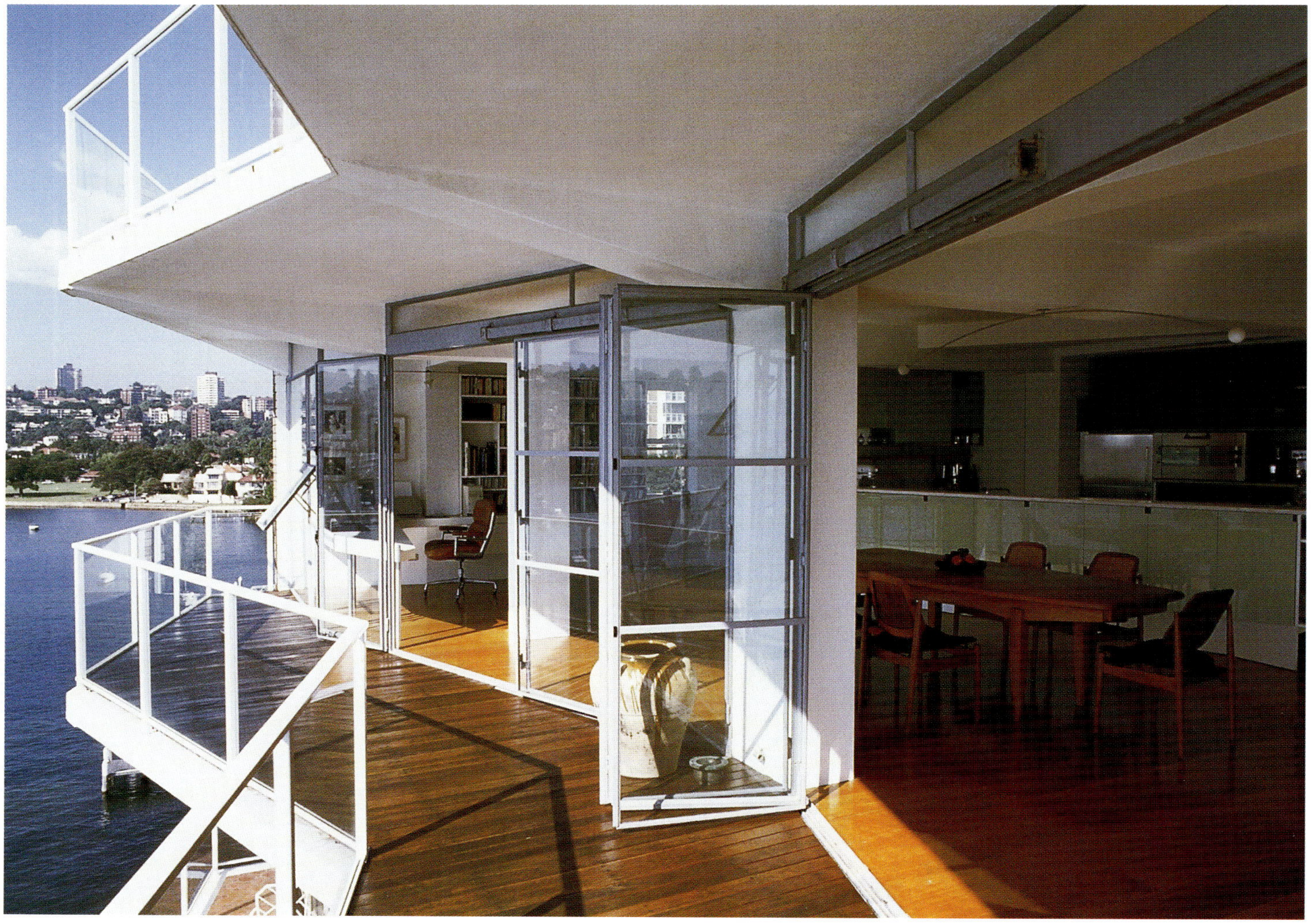

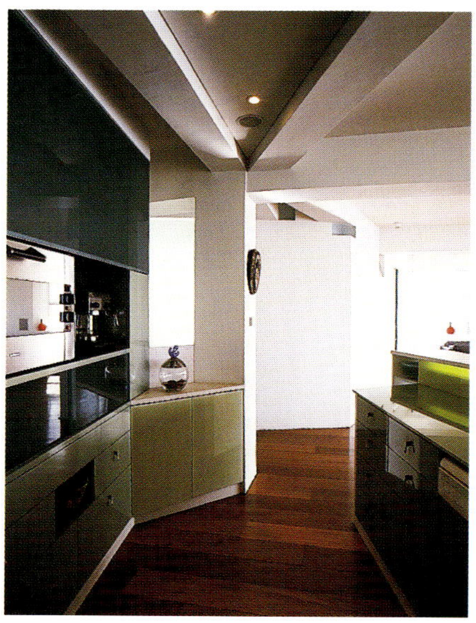

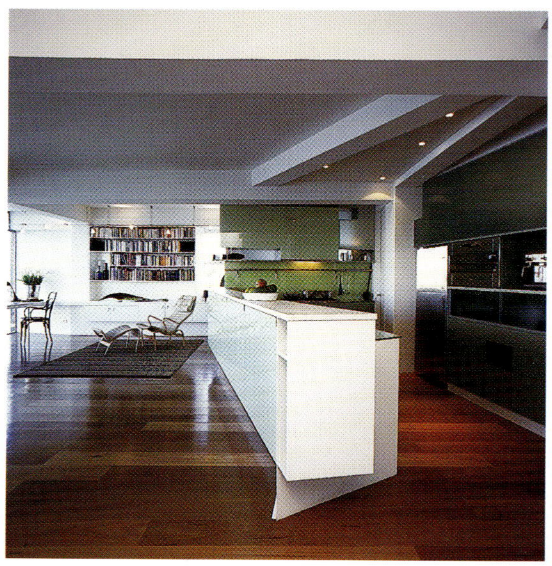

TOP: BALCONY BOTTOM LEFT: KITCHEN BOTTOM RIGHT: LIVING AREA AND KITCHEN OPPOSITE TOP: KITCHEN AND ENTRY OPPOSITE BOTTOM: KITCHEN AND LIVING AREA

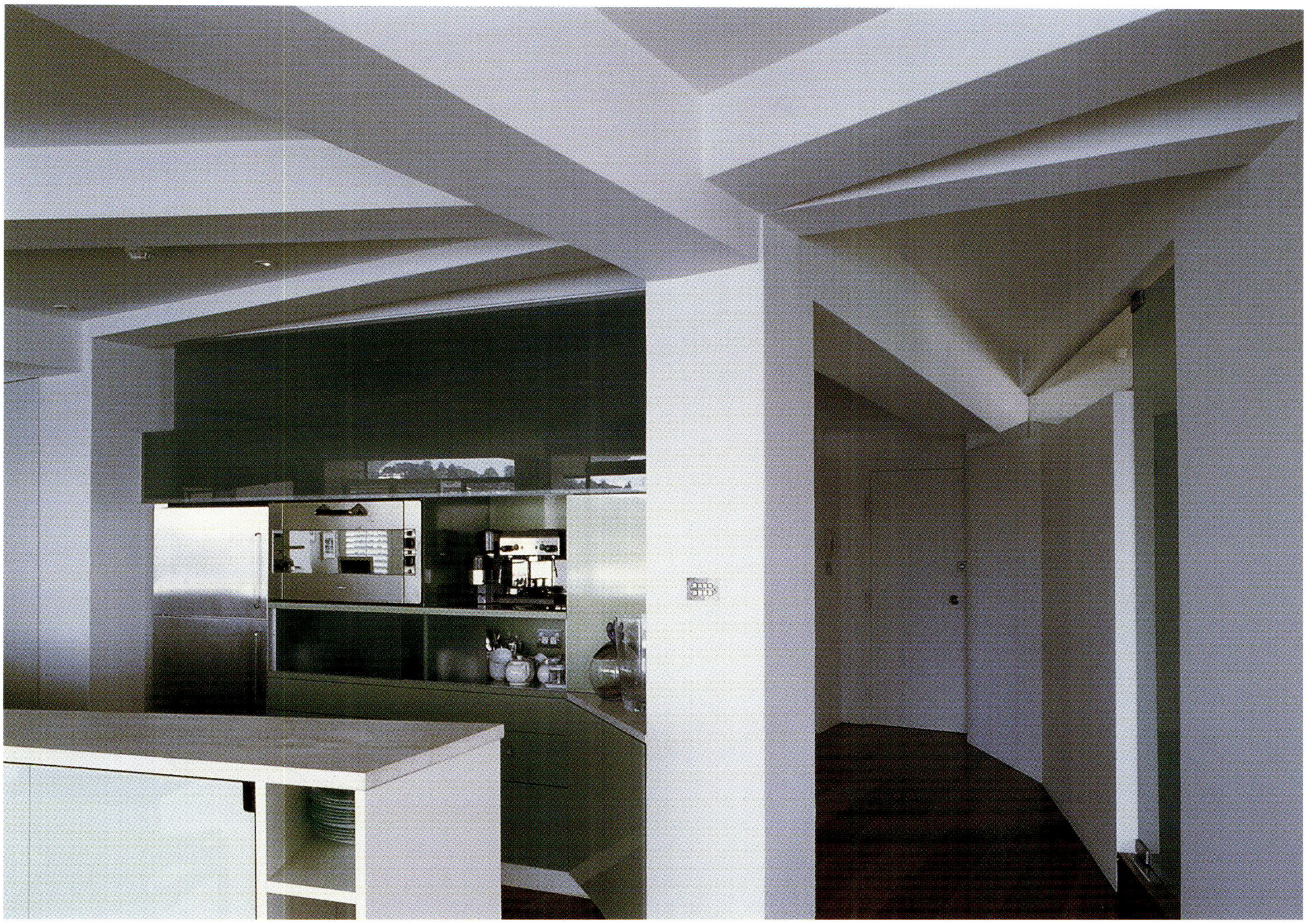

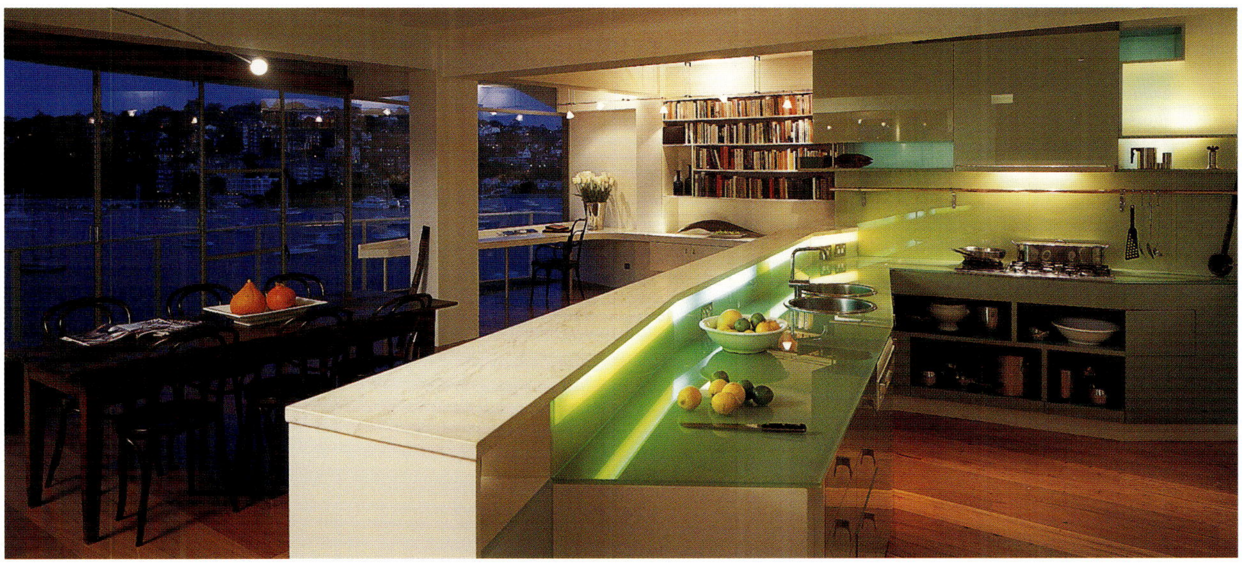

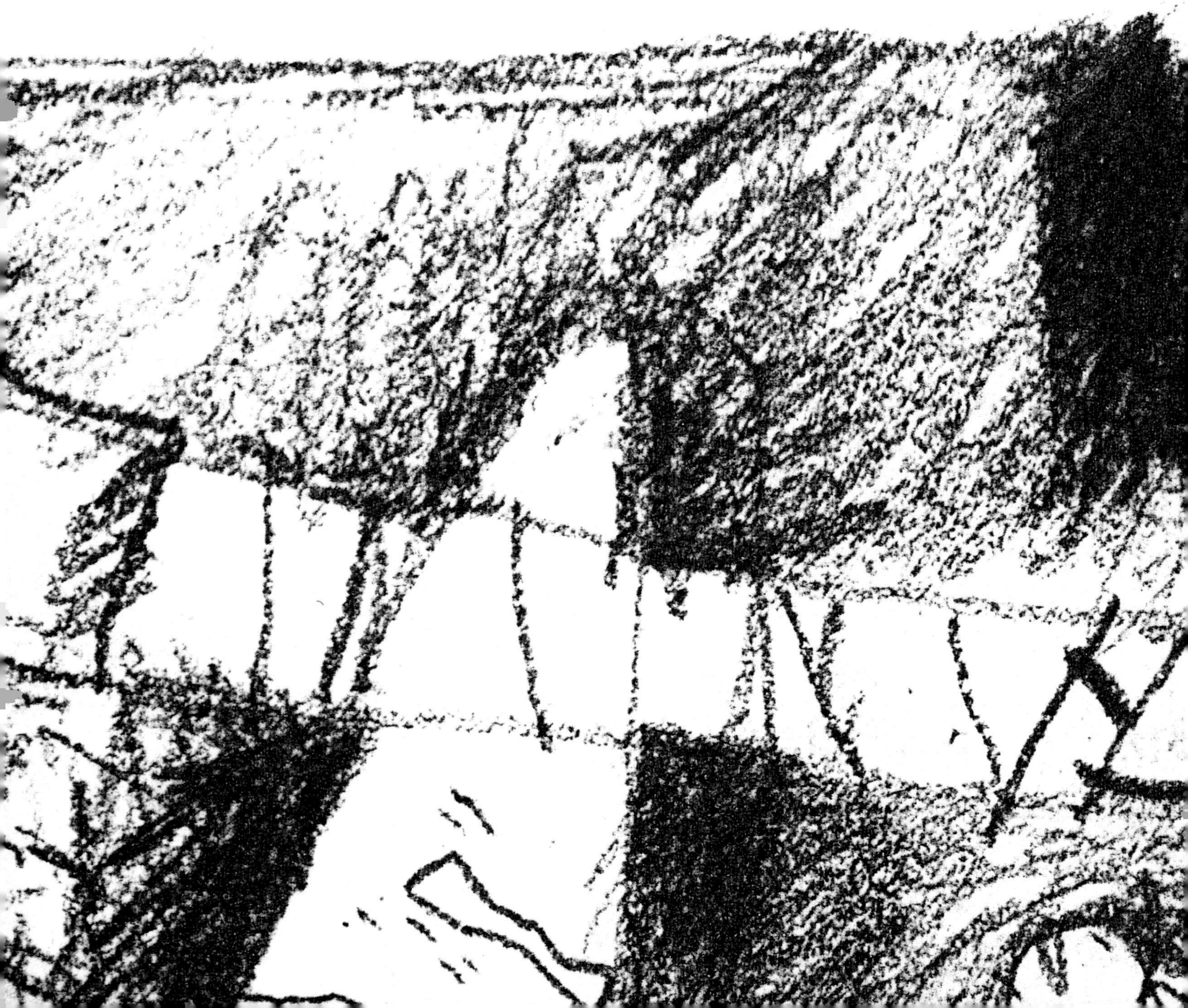

PROJECT SIX

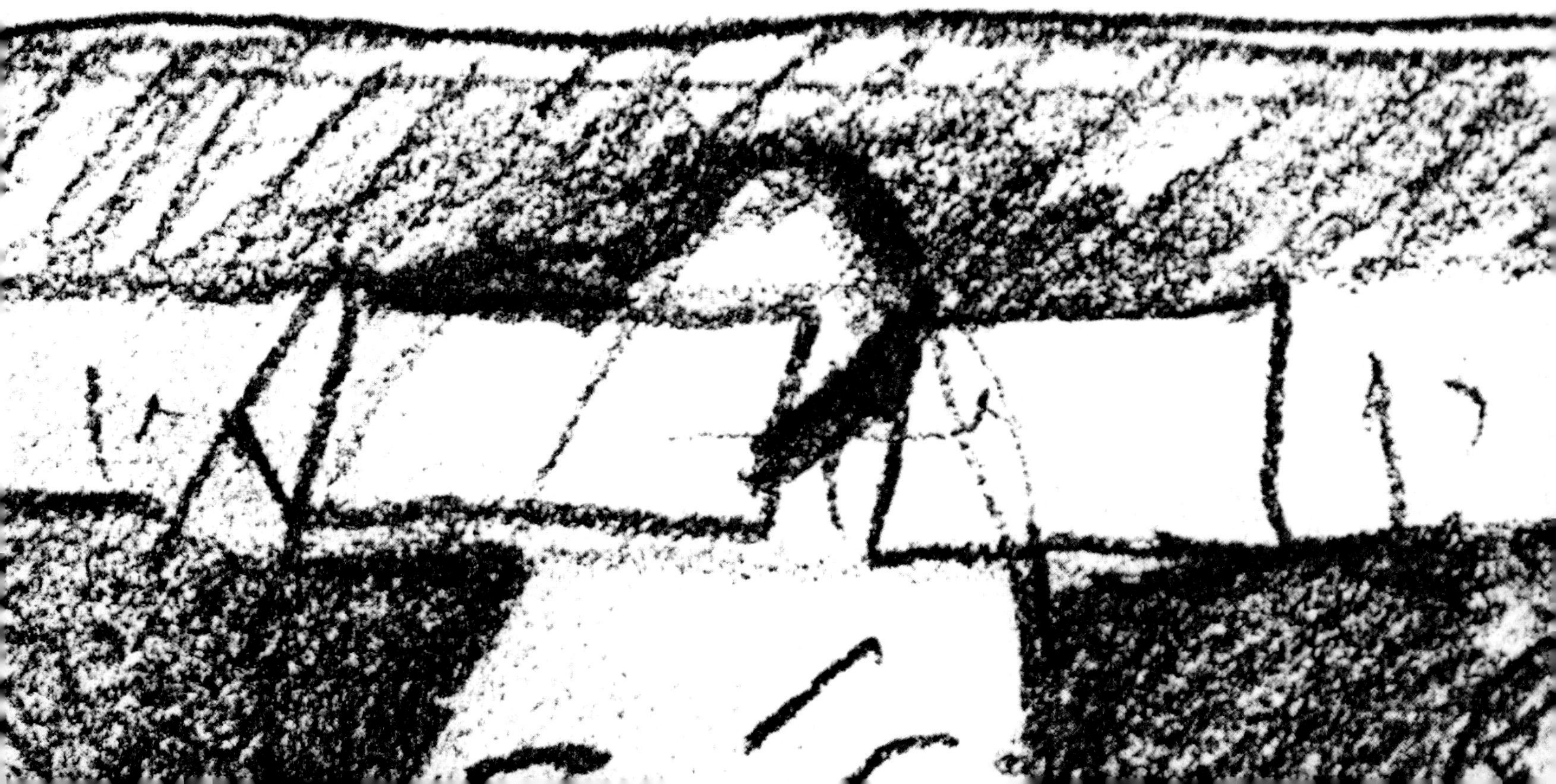

HILLSIDE HOUSE VAUCLUSE, SYDNEY

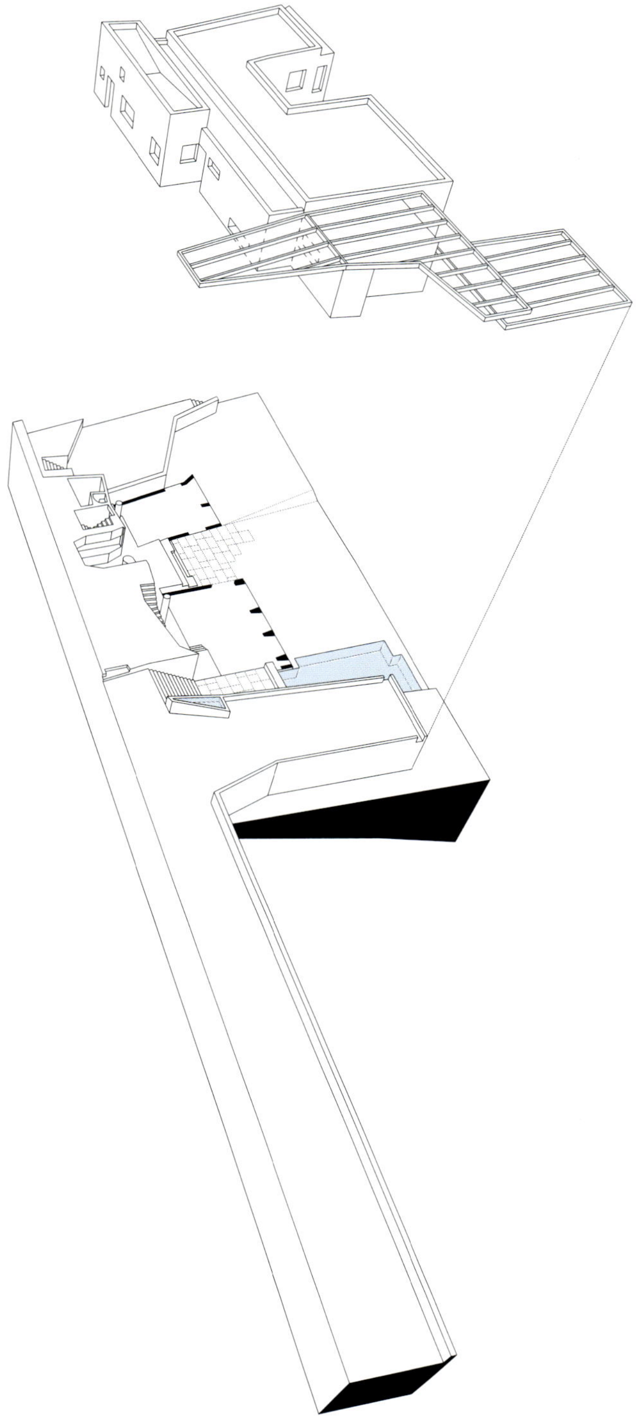

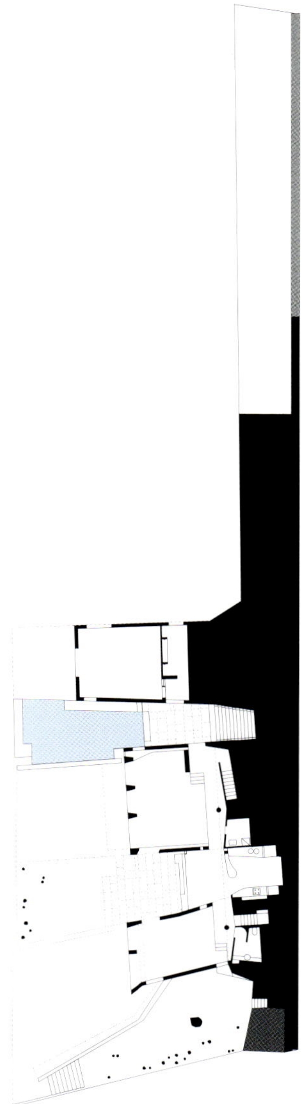

PROJECT TEAM Neil Durbach, Camilla Block, Lisa Le Van, Charles Chong, David Ostinga, Emma Williamson Date of completion 1997

This house is an exposition of the site itself: a long approach against a stone wall crossed by a plate extending to the harbour views. The site framed the key questions, which in turn informed the strategy. Could a courtyard house become as pleasurable as the tennis court site it once occupied? Could a linear system be eased into a striated system running both perpendicular and parallel to the view? Could a difficult height covenant be reconciled with the expense and waste of excavation?

The house is mounted about a crumpled gallery which is crossed by two courts locking across the site and revealing its section – one over, one under. Each court will one day be a pool or a copse. At right angles, and flush with the raised first floor, is a small north-facing garden floating between the stone retaining wall and the compressed volume of the house. Parallel to, but terminating, the banded series of courts and rooms is a colossal Moreton Bay fig tree with its shaded and stepped garden edging the site. Attached to the spatially complex gallery spine are the large generic rooms, gently flared and quite calm.

GALLERY AND STAIR

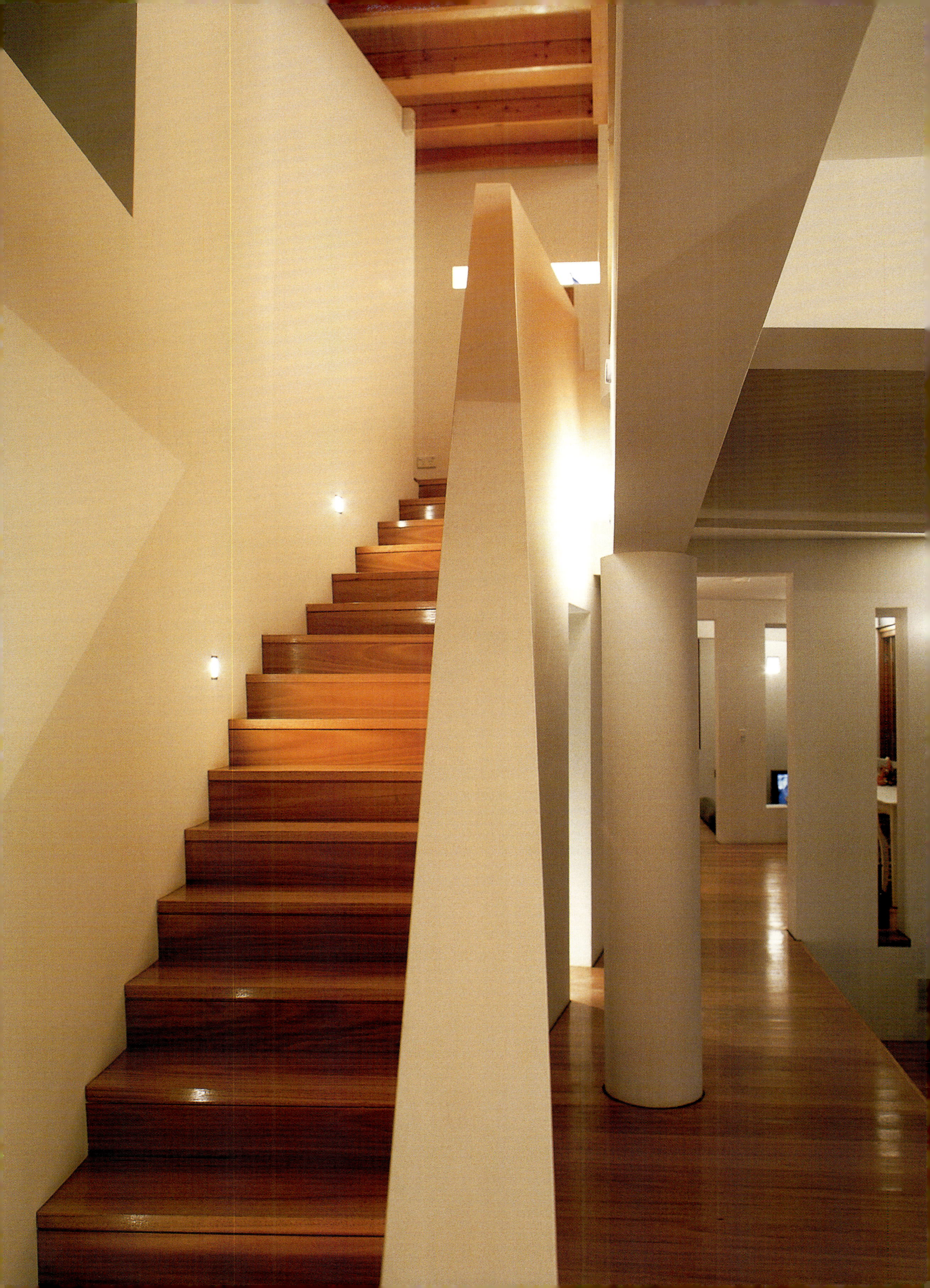

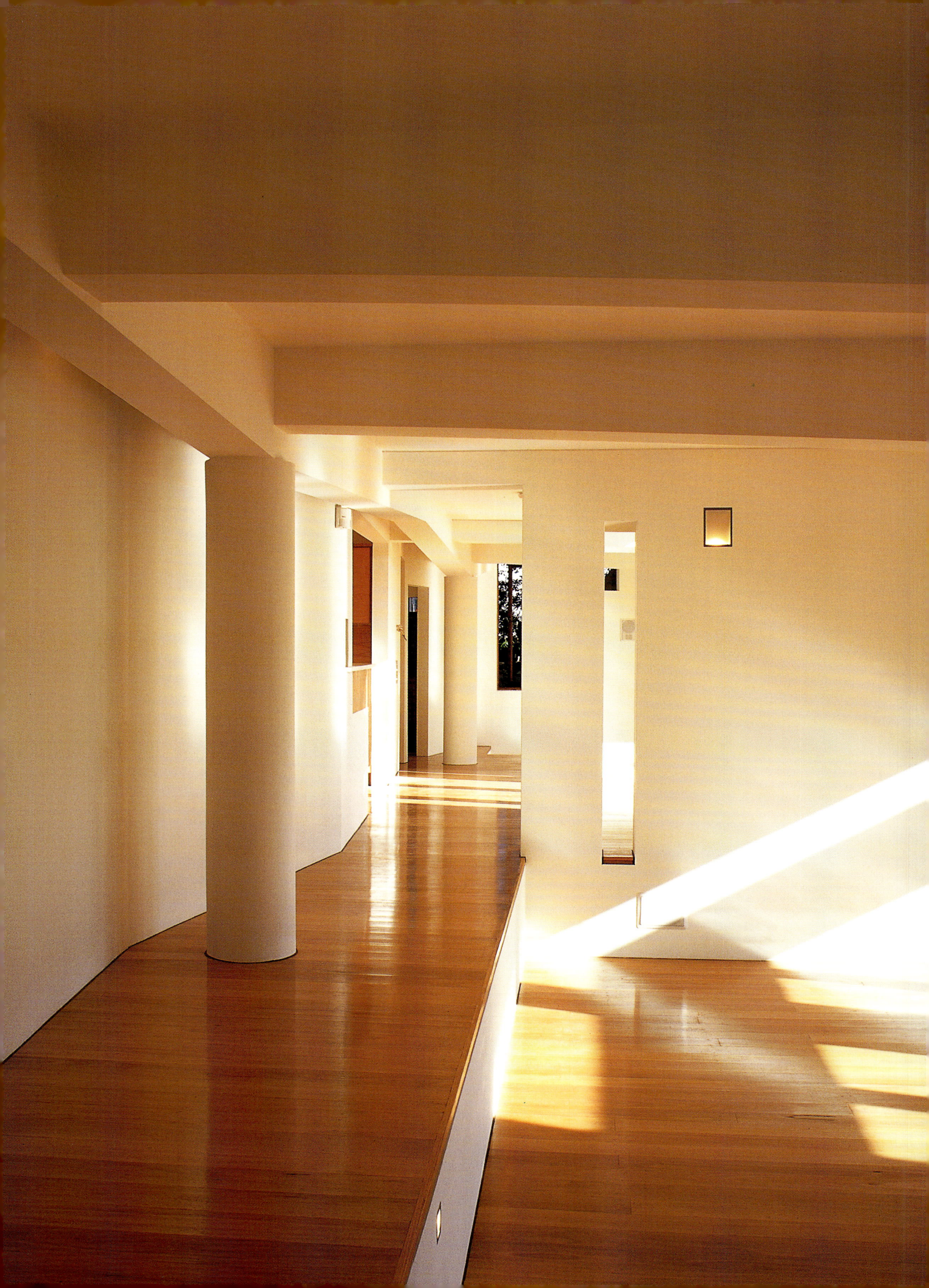

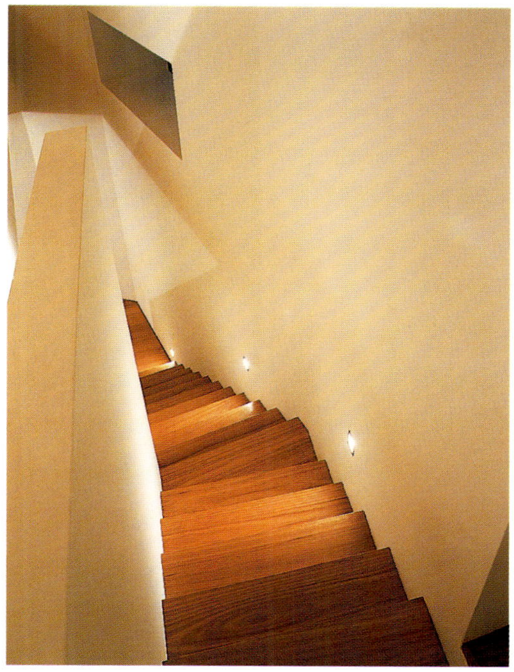

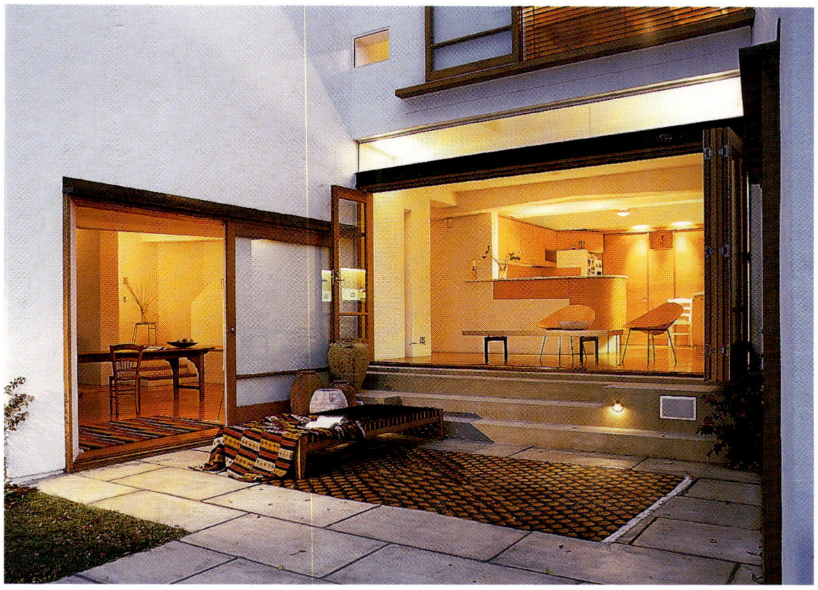

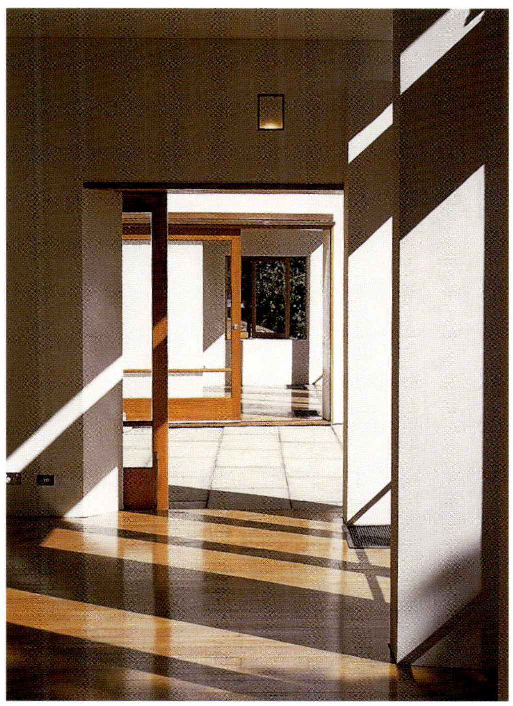

LEFT: GALLERY AND LIVING ROOM CENTRE: COURTYARD AND KITCHEN TOP: STAIR FROM FIRST FLOOR BOTTOM: LIVING ROOM THROUGH COURTYARD TO FAMILY ROOM

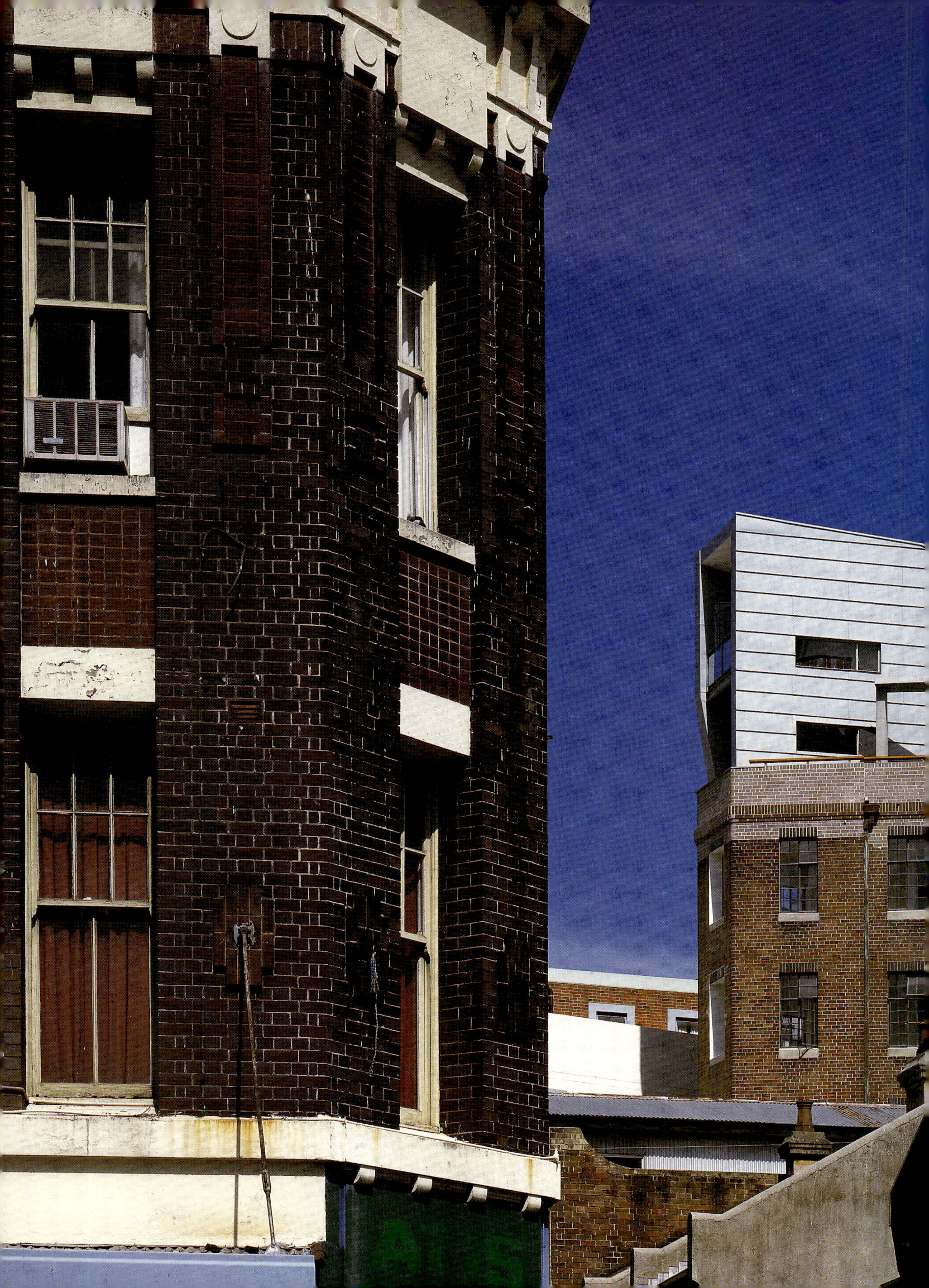

PROJECT SEVEN

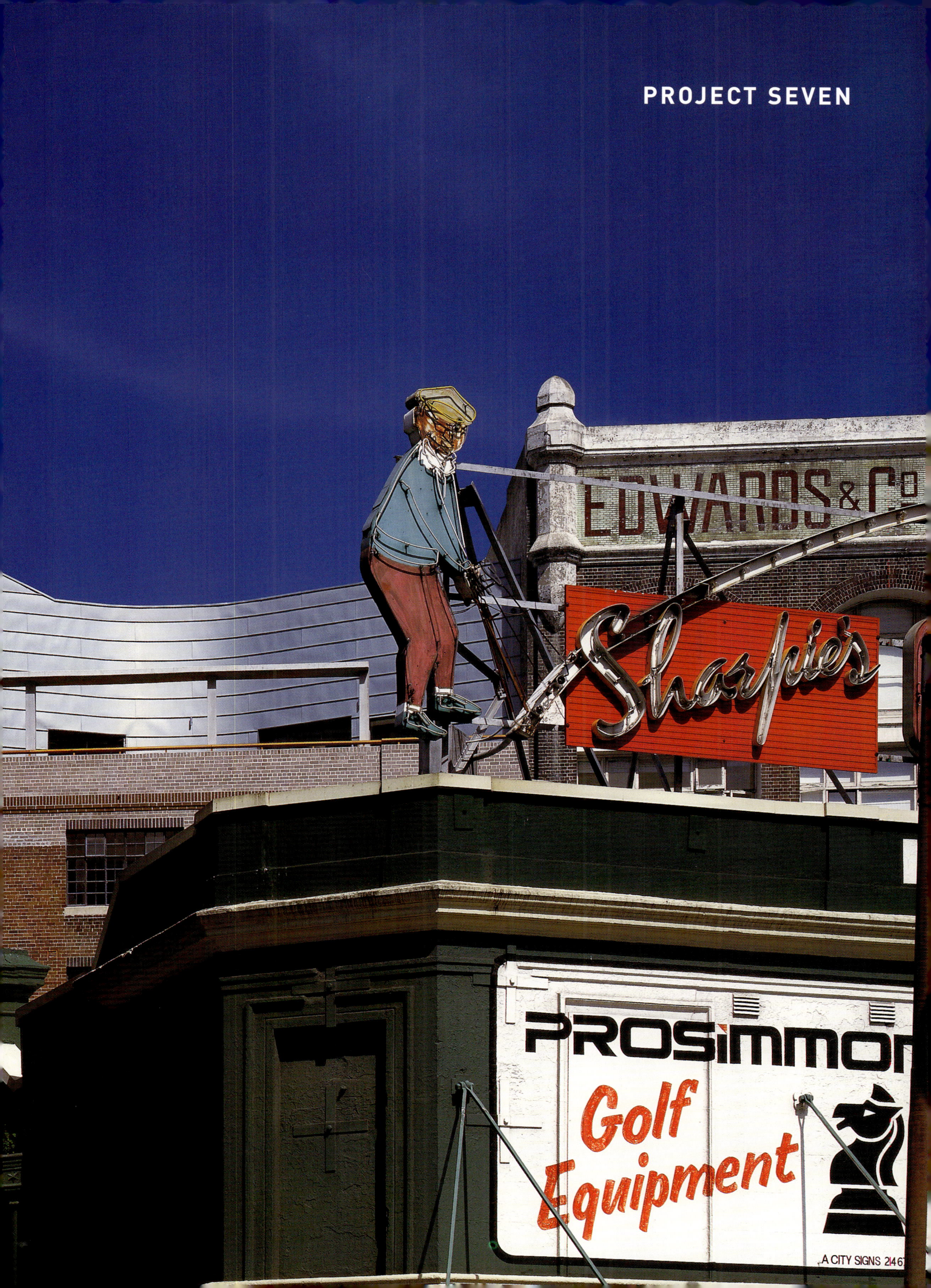

FOSTER STREET APARTMENT SURRY HILLS, SYDNEY

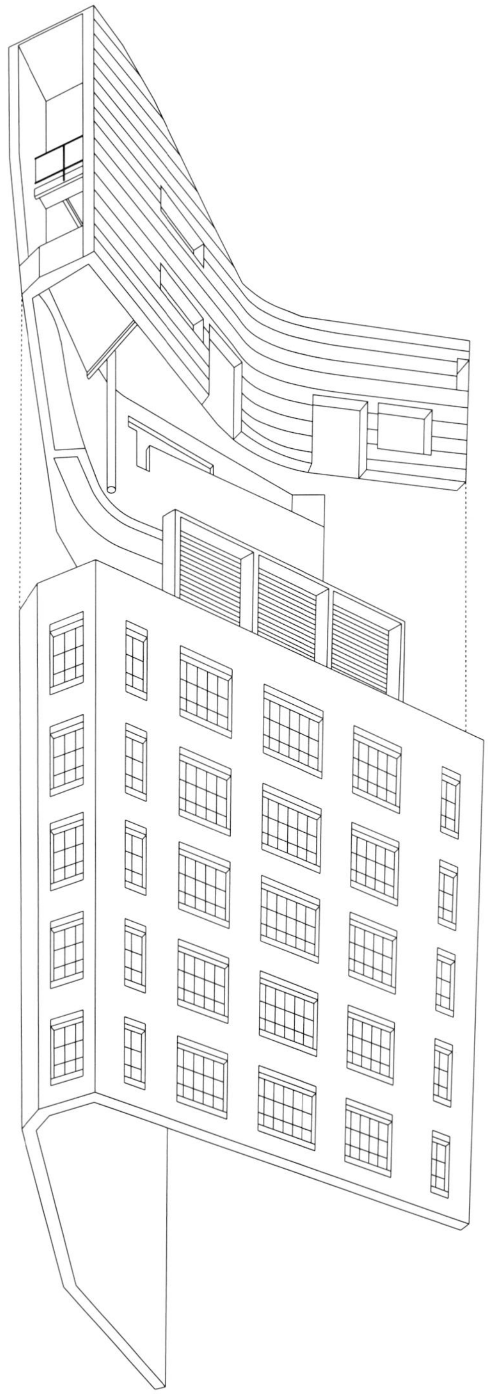

In the hilly and disordered end of town, this penthouse sits on top of a five storey triangular brick warehouse. The precise perimeter of the roof top site enables a plastic response. A stable and solid base becomes an uncertain and slippery top.

Emerging from the hinged corner are three splaying strips. The first band holds domestic and service spaces, edged by a curved wall that thickens for a concealed stair. The second is a bending and tapering living space that opens selectively side to side, and rises up to the northern raised kitchen, to hold the city centre in its gaze. The third is a terrace with a curved zinc-clad back overlooking the city. This curved court completes a subtle urban pattern of an oval hollow, inscribed by a foreground of low buildings, before the dramatic towers of the city rise upward. A brise soleil, designed to house timber shutters, turns the harsh western sun to its advantage.

The form of the penthouse is intentionally mute and sculptural, a tough response to a tough neighbourhood. This muscular exterior belies the smooth, calm and elastic interior, a distinctive mix of urban immersion and overview.

PROJECT TEAM Neil Durbach, Camilla Block, Joe Grech, Lisa Le Van

Date of completion 1997

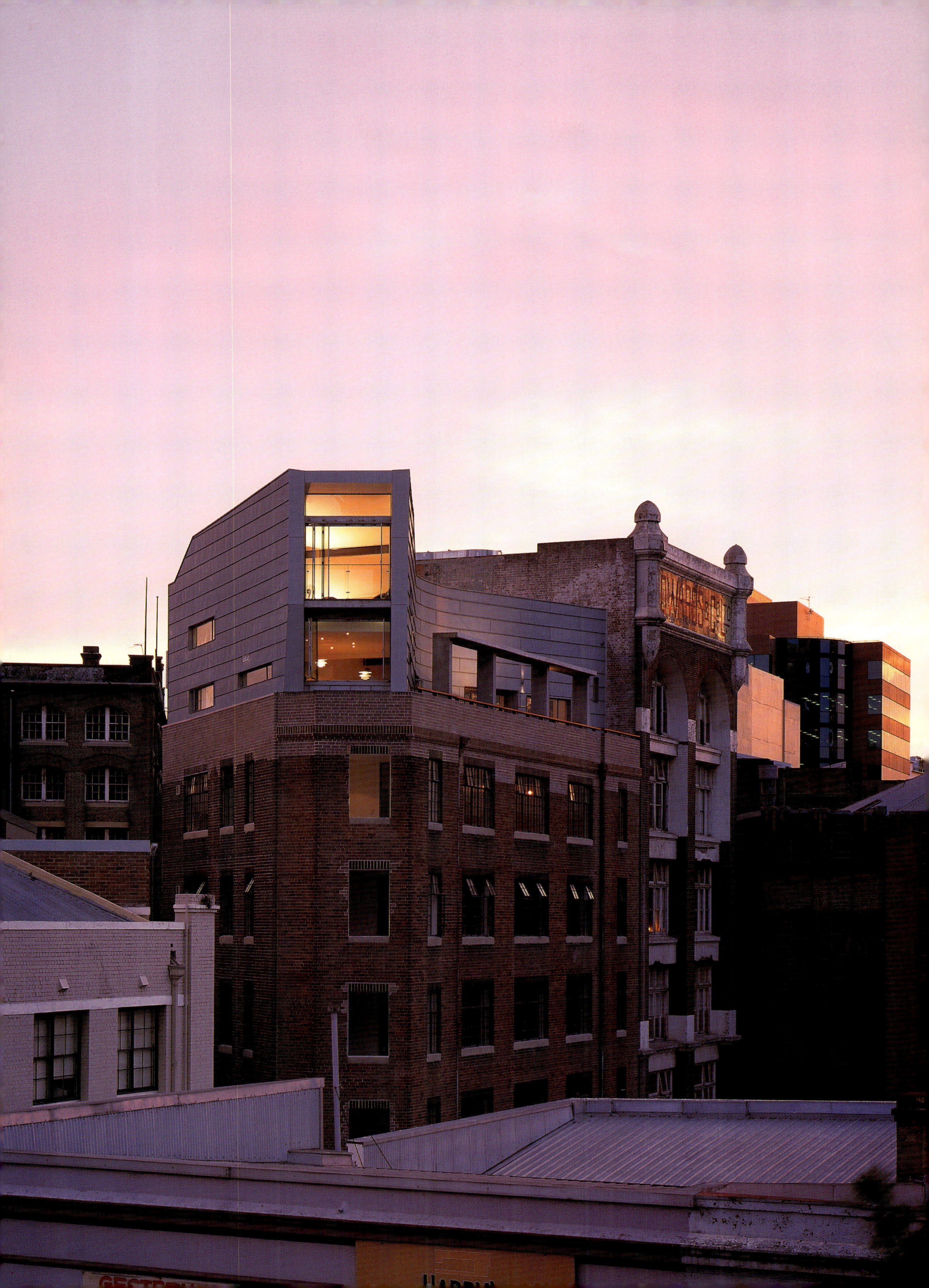

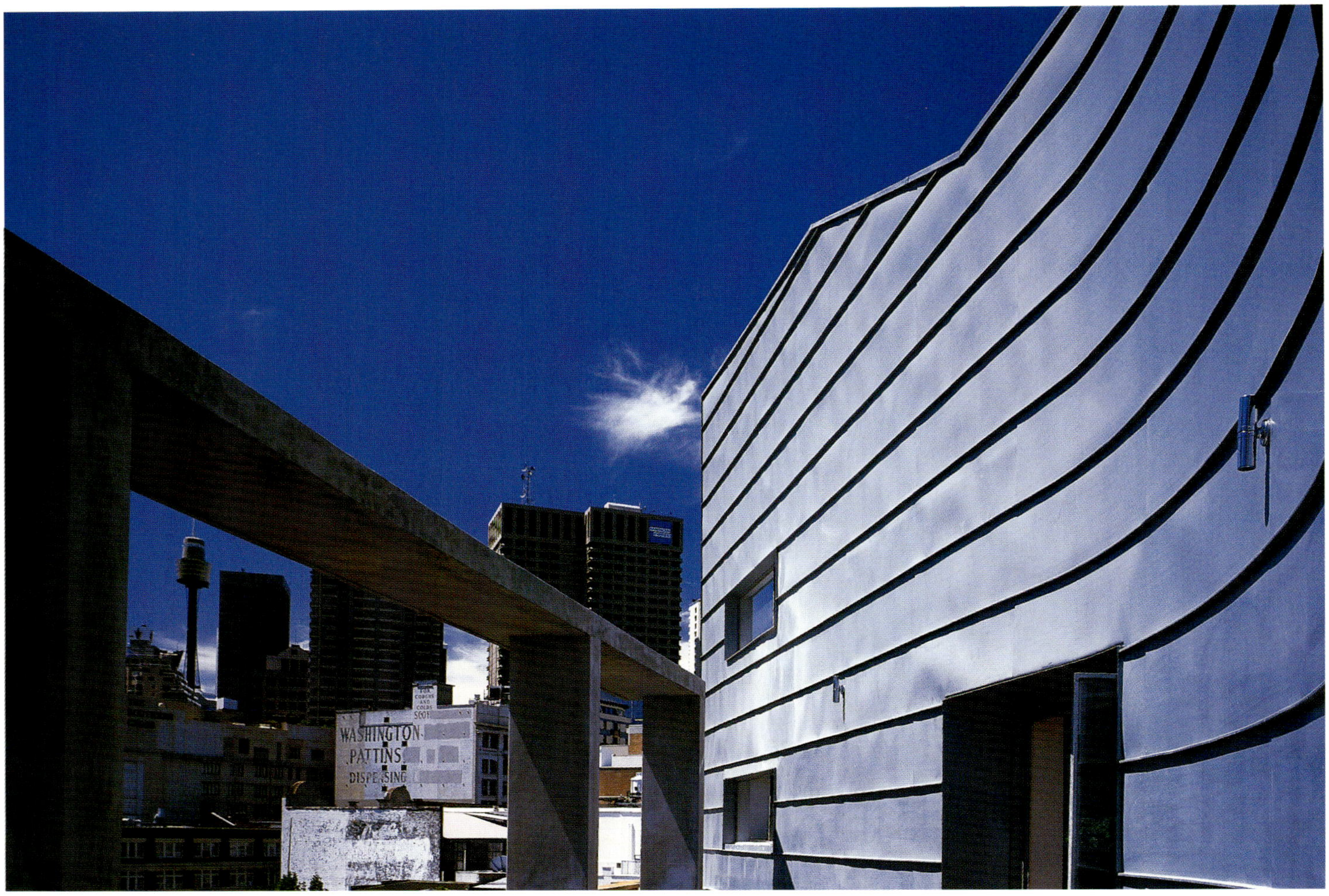

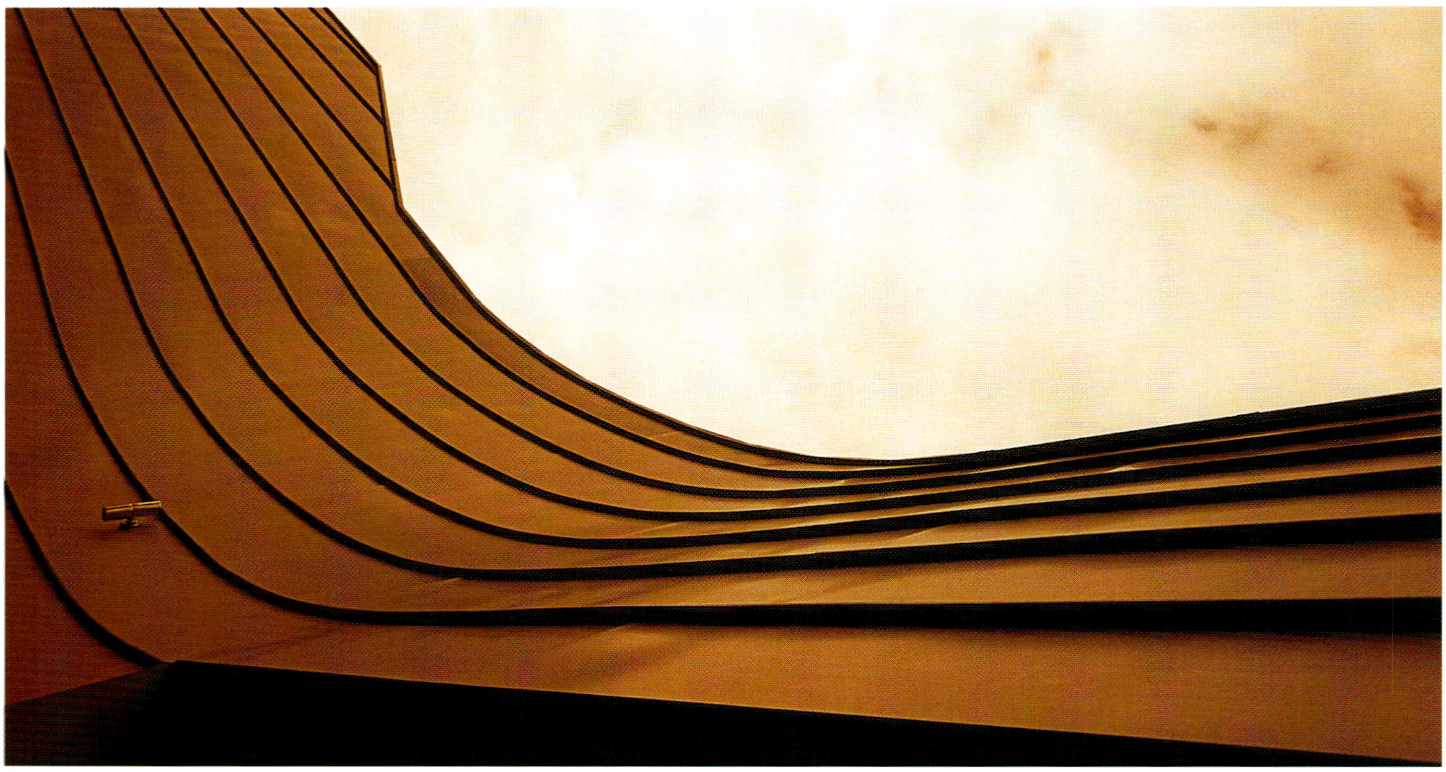

TOP: BRISE SOLEIL AND WESTERN FACADE BOTTOM: CURVED ZINC WALL OPPOSITE: TERRACE

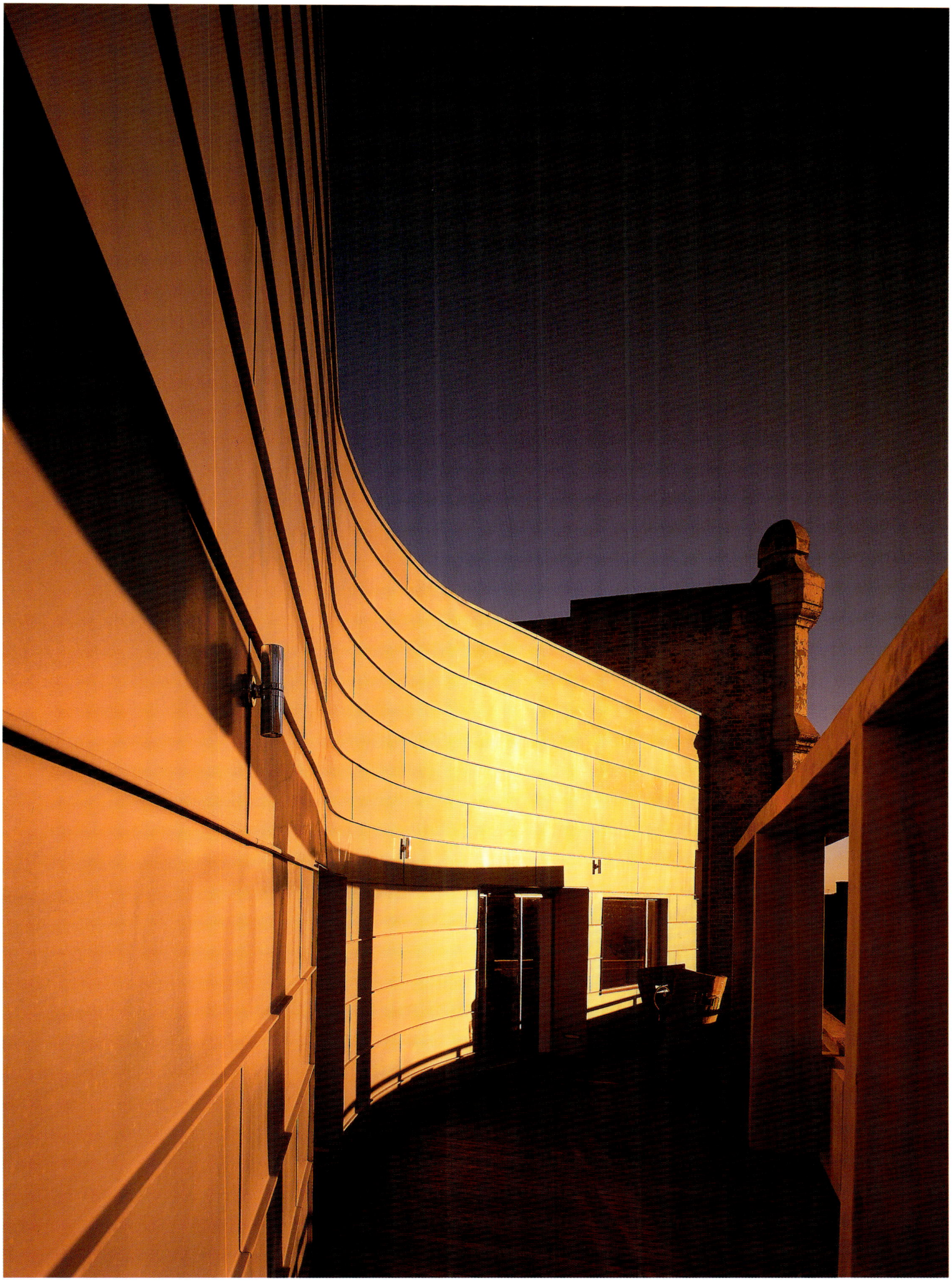

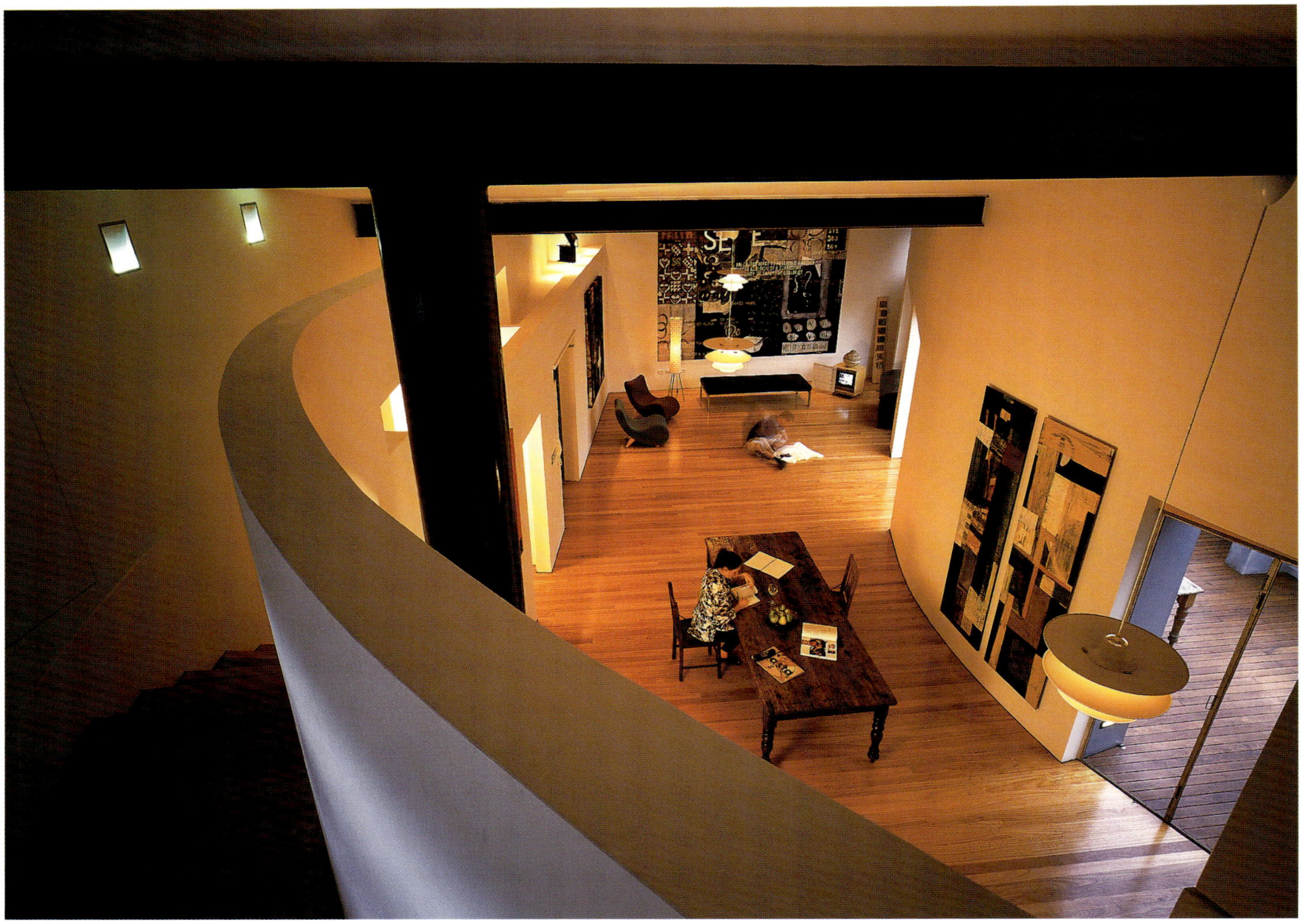

TOP: LIVING AREA FROM GALLERY BOTTOM: STAIR FROM GALLERY OPPOSITE: GROUND FLOOR ENTRY LOBBY

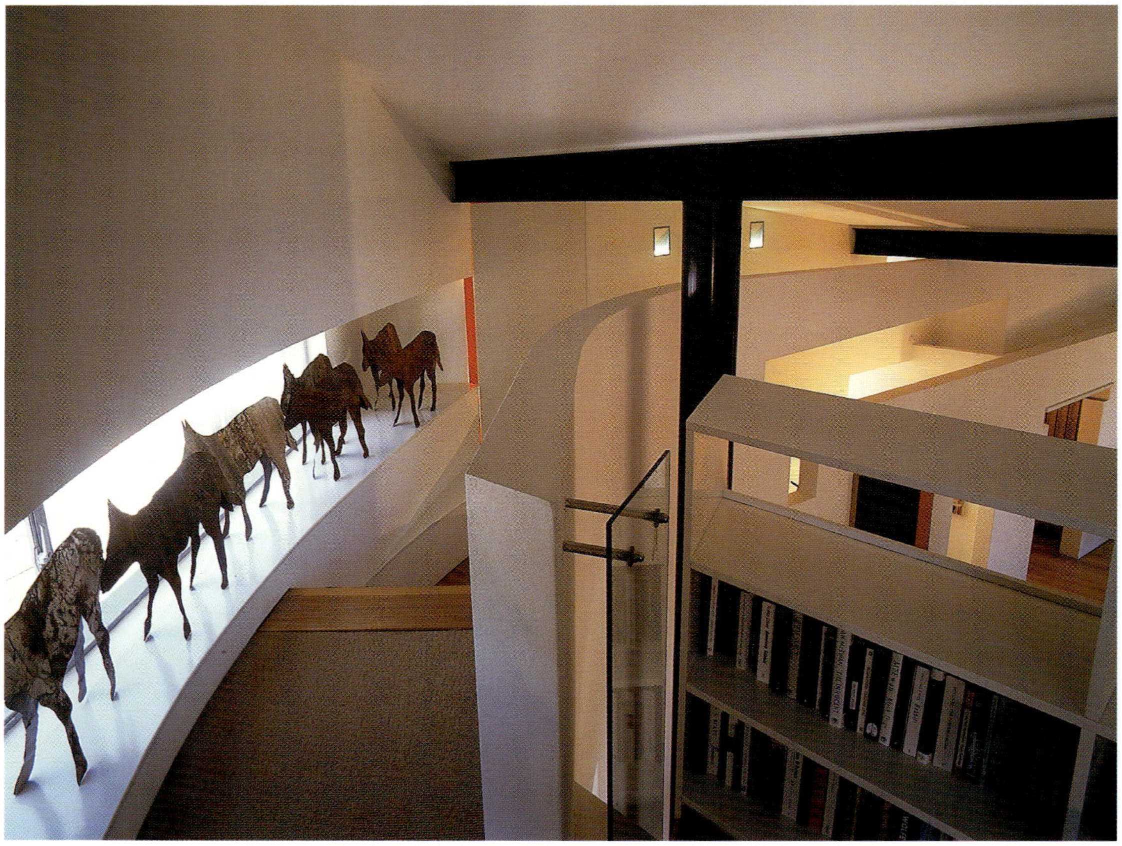

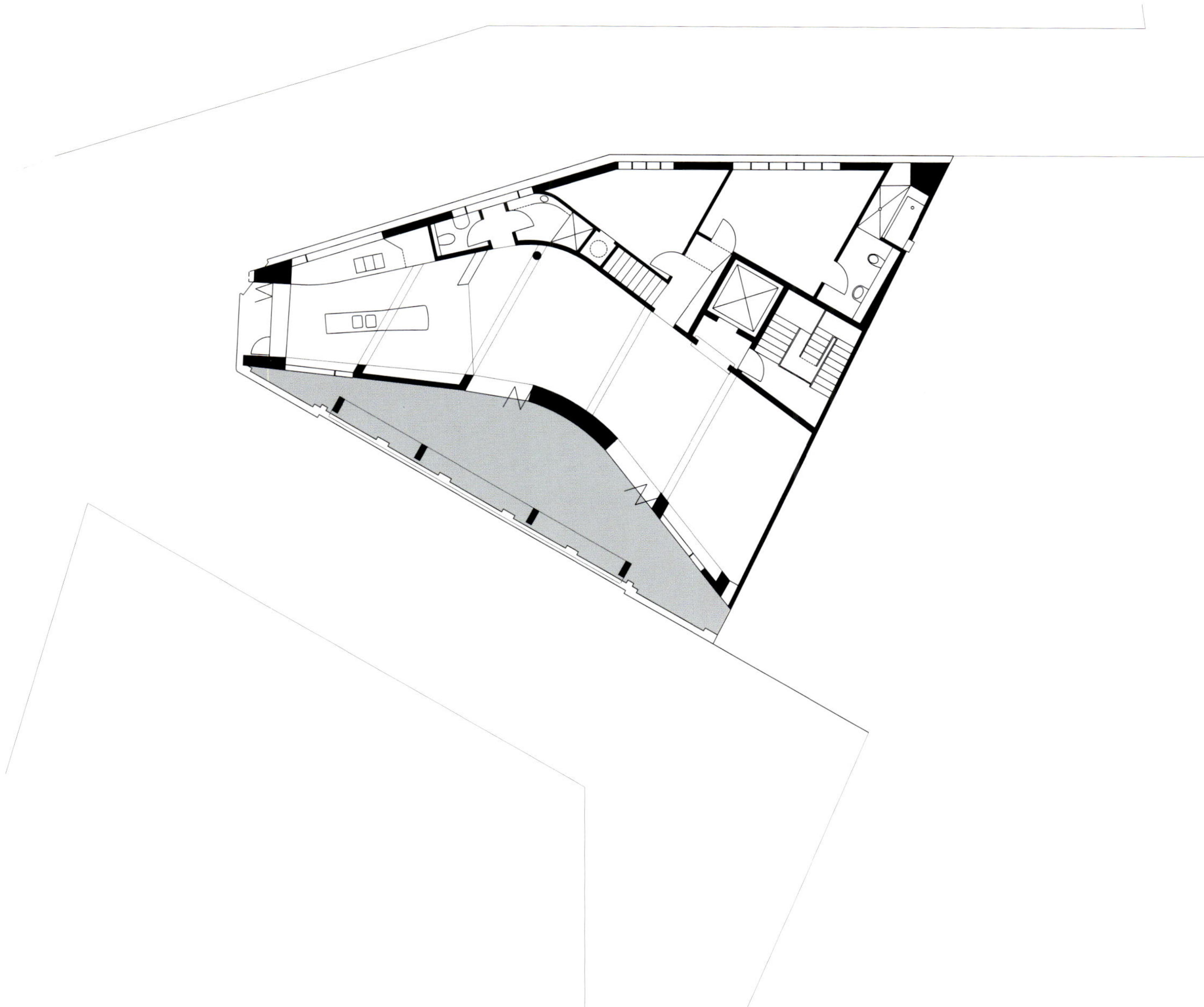

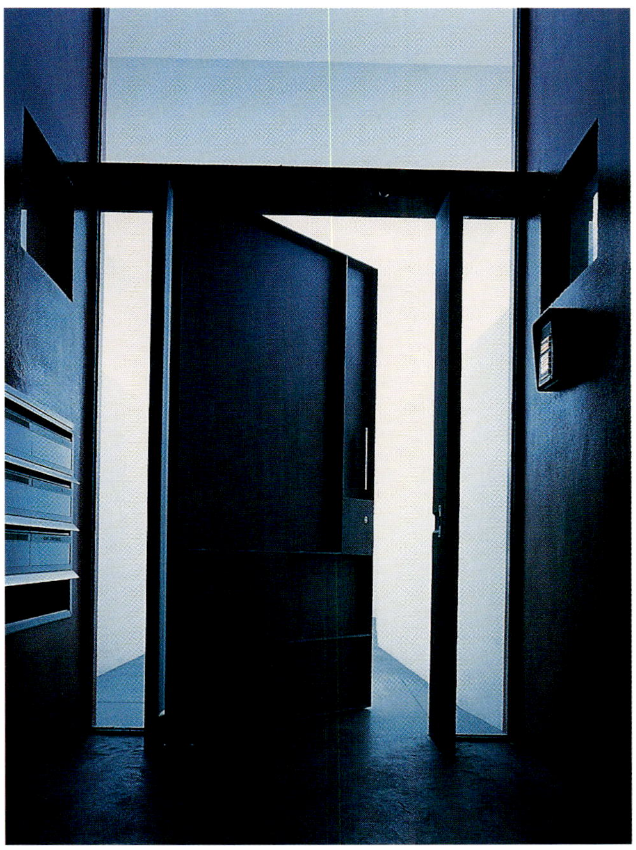

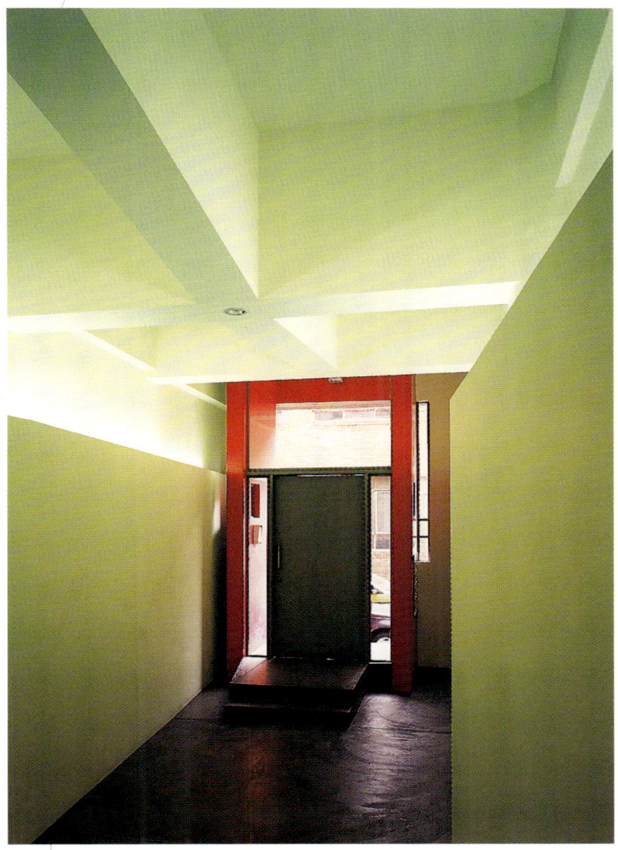

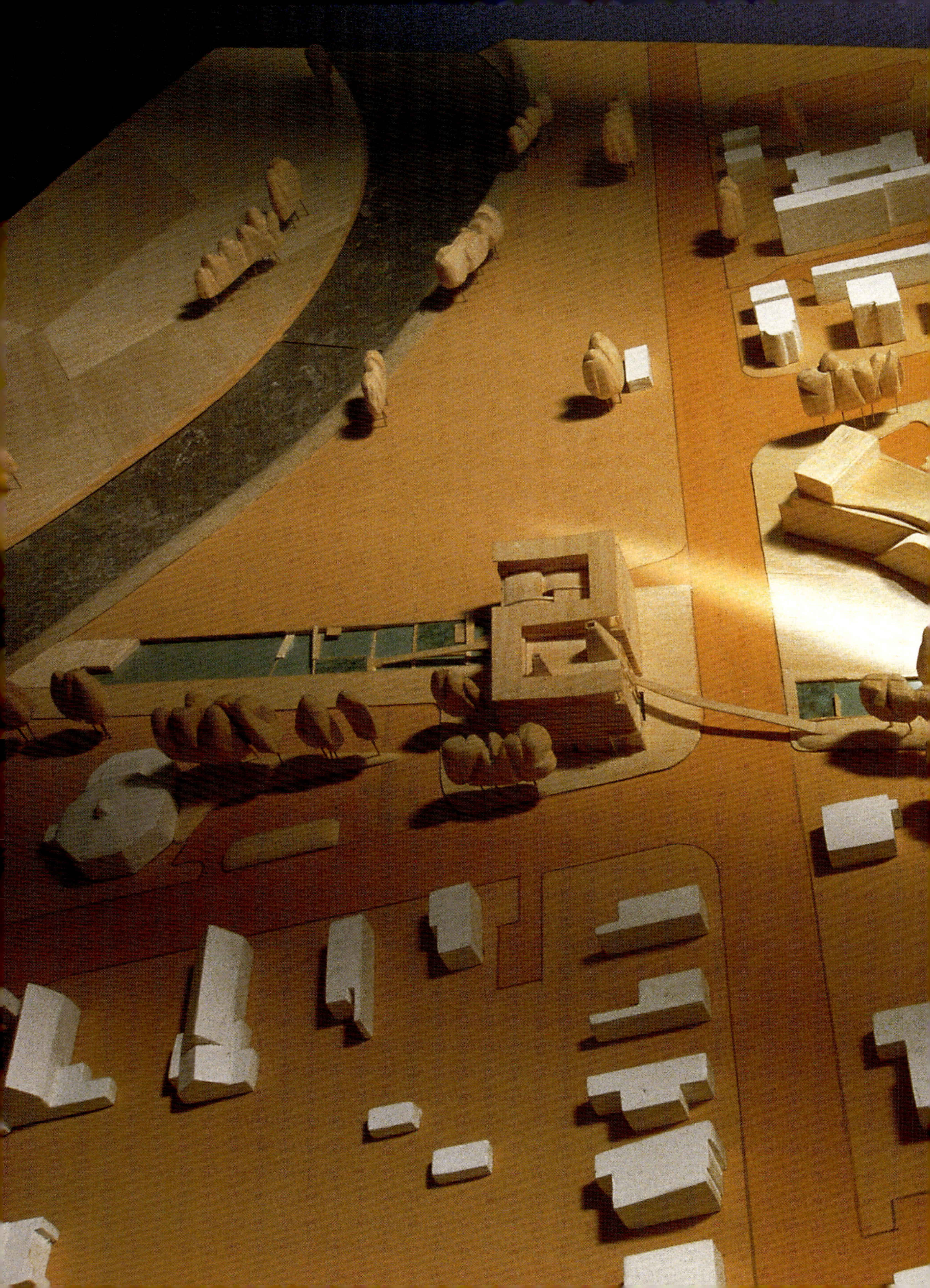

PROJECT EIGHT

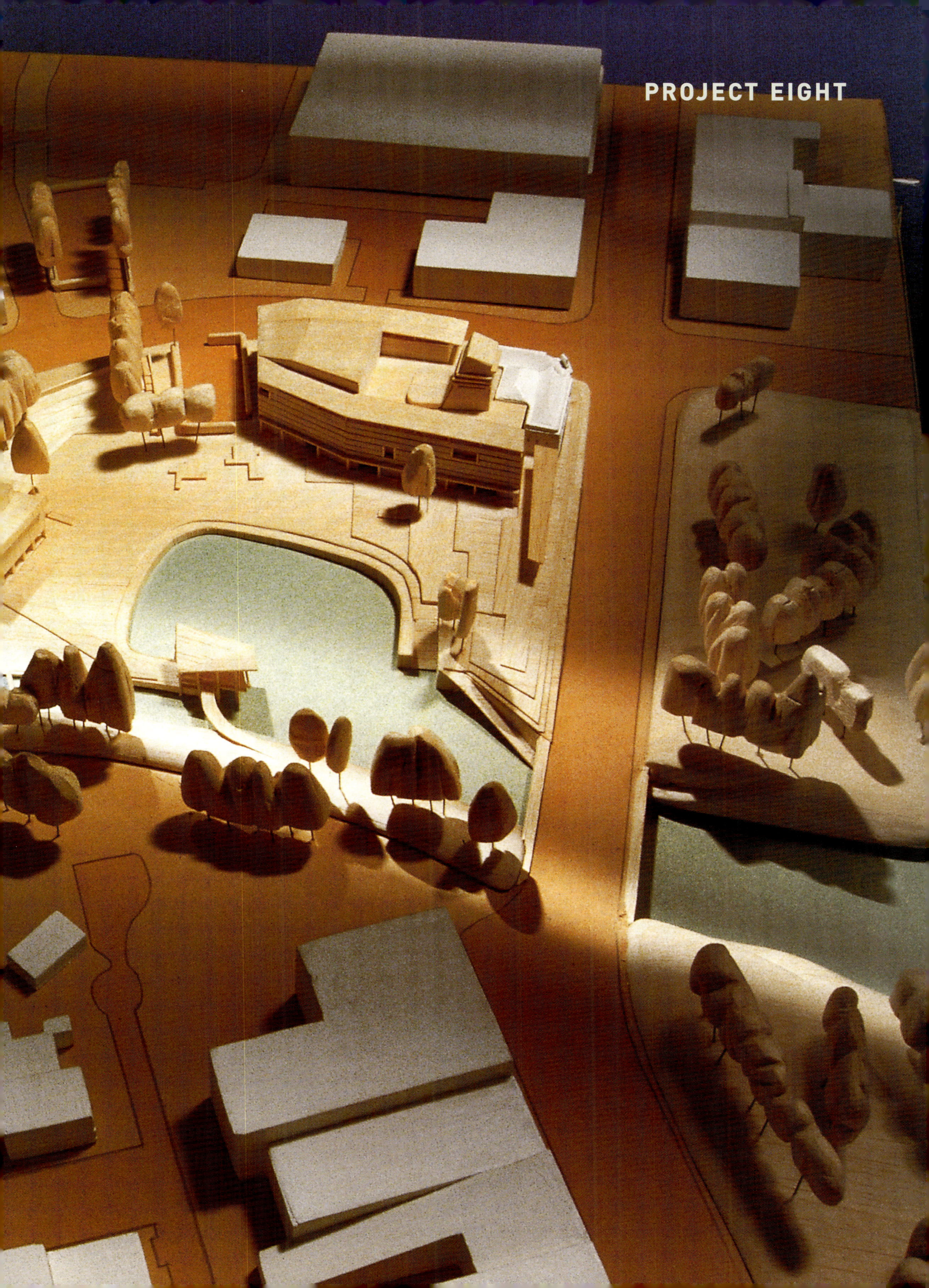

WAGGA WAGGA CIVIC CENTRE

PROJECT TEAM Neil Durbach, Camilla Block, Stephen Collier, Lisa Le Van, Emma Williamson, Ian Perlman 1995

A swathe of public land currently cuts through the city grids, leads to, but never reaches the Murrumbidgee River. Centred on the site is the contained, still water of an existing lagoon, contrasting with the broad flowing passage of the river. The subject of an international competition, the site housed an existing theatre and historic council chamber, loosely gathered around the lagoon. The brief called for a new council chamber and offices, a gallery and library

The proposal for a collection of new and reworked civic buildings set out to reorient the site toward the river, and, through a set of buildings, into a representative urbanity. The scheme explored an architecture of civic synthesis. Specific buildings and related landscapes identify the collective and the singular conditions of civic life in an Australian town. The Memorial Gardens are loosely continued into the site. The near-radial character of the land and water gardens acknowledges the lagoon as a gathering point before the river, without freezing it into a metaphoric end. The two buildings housing collective functions, the council chambers and theatre, orient towards this centre, creating a place of urban intensity.

The perimeter of offices of the historic council offices building gives

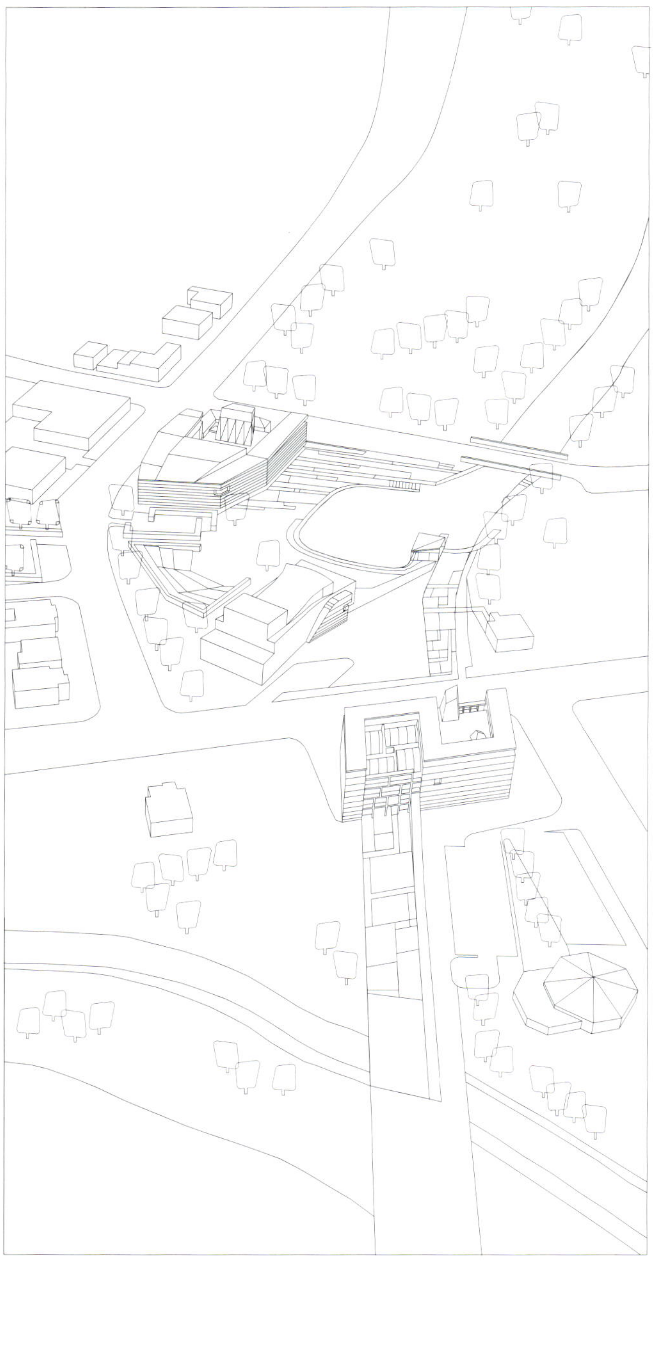

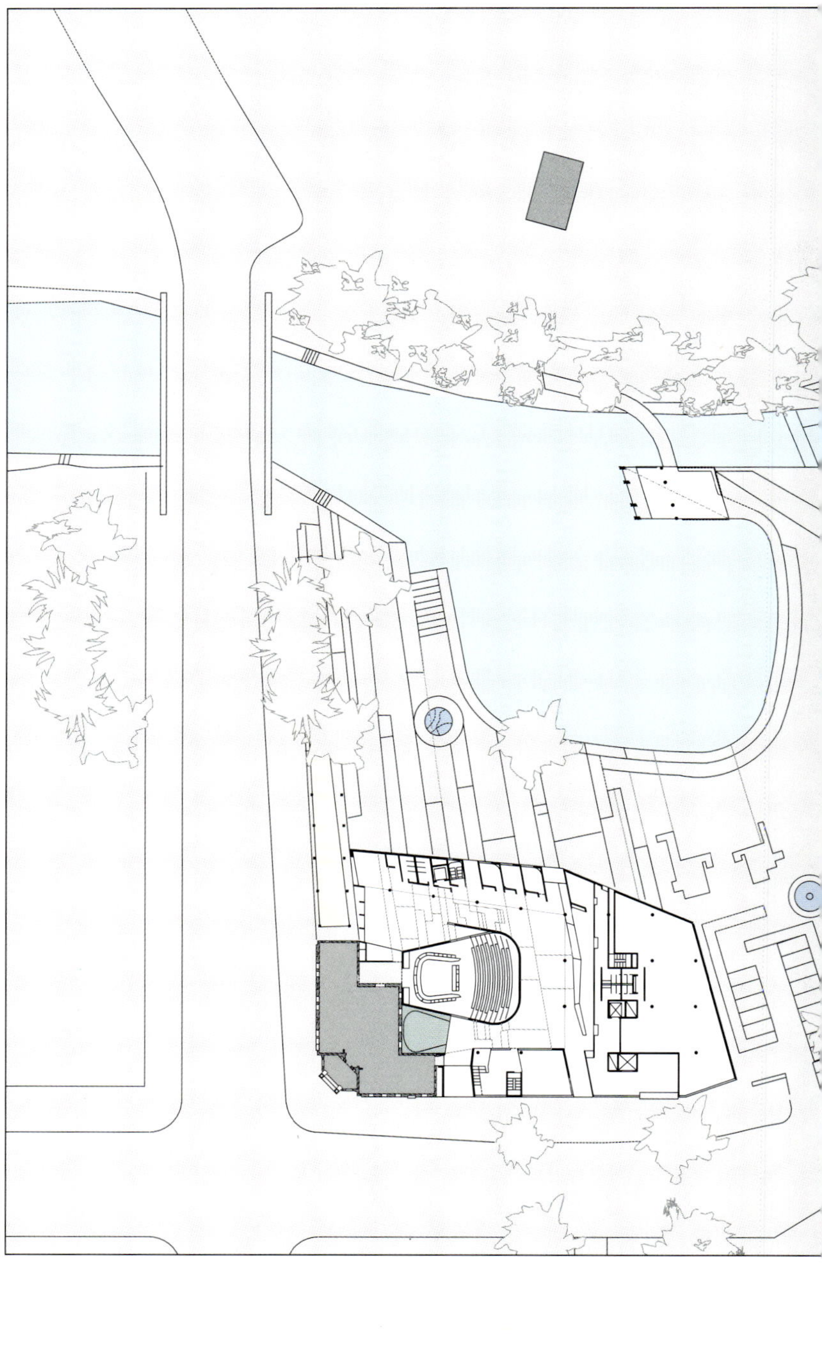

the unseen council chamber a presence on the street. The proposal completes the street building, adding its own string of office spaces around the new council chamber. A new civic lantern celebrates the unseen interior on the main street.

In the civic theatre, a pragmatic reconfiguration is achieved by maintaining the lateral building envelope and refolding the roof to accommodate new balcony seating, served by two slim staircase envelopes.

Bermed, stepped and wooded, the gardens serve to intensify the edges of this space, focussing on the remade lagoon. New water gardens reach to the janus-like library and gallery and onto the river beyond. The building links the townscape with the river. The gallery, on the town side, adds its own cluster of steeple skylights to the sky-piercing collection on the point. The library addresses the open fields on the opposite bank from a pinwheel reading room and rolling roof. The winding circulation route evokes the bending of the river, which mediates between the town and the open land beyond.

Separate buildings and linked gardens and water gardens, the town begins. Separate buildings within a binding landscape transcend any single structure: they becomes a legible town centre. The eastward orientation of the scheme, bending towards the curve in the river, gathers together the new and the existing. This connection establishes the condition of the site. Three buildings: an interlocking case – the council administration offices, the unpeeled proto-modernist theatre, and the dual-faced library and gallery all recall a town on the river.

LANDSCAPING

The memorial gardens are loosely continued into the site along the poplar and red rivergum-lined water gardens and the heavily wooded berm adjacent to the civic theatre.

BUILDINGS

In this way disparate elements are stitched together into a single arcadian identity.

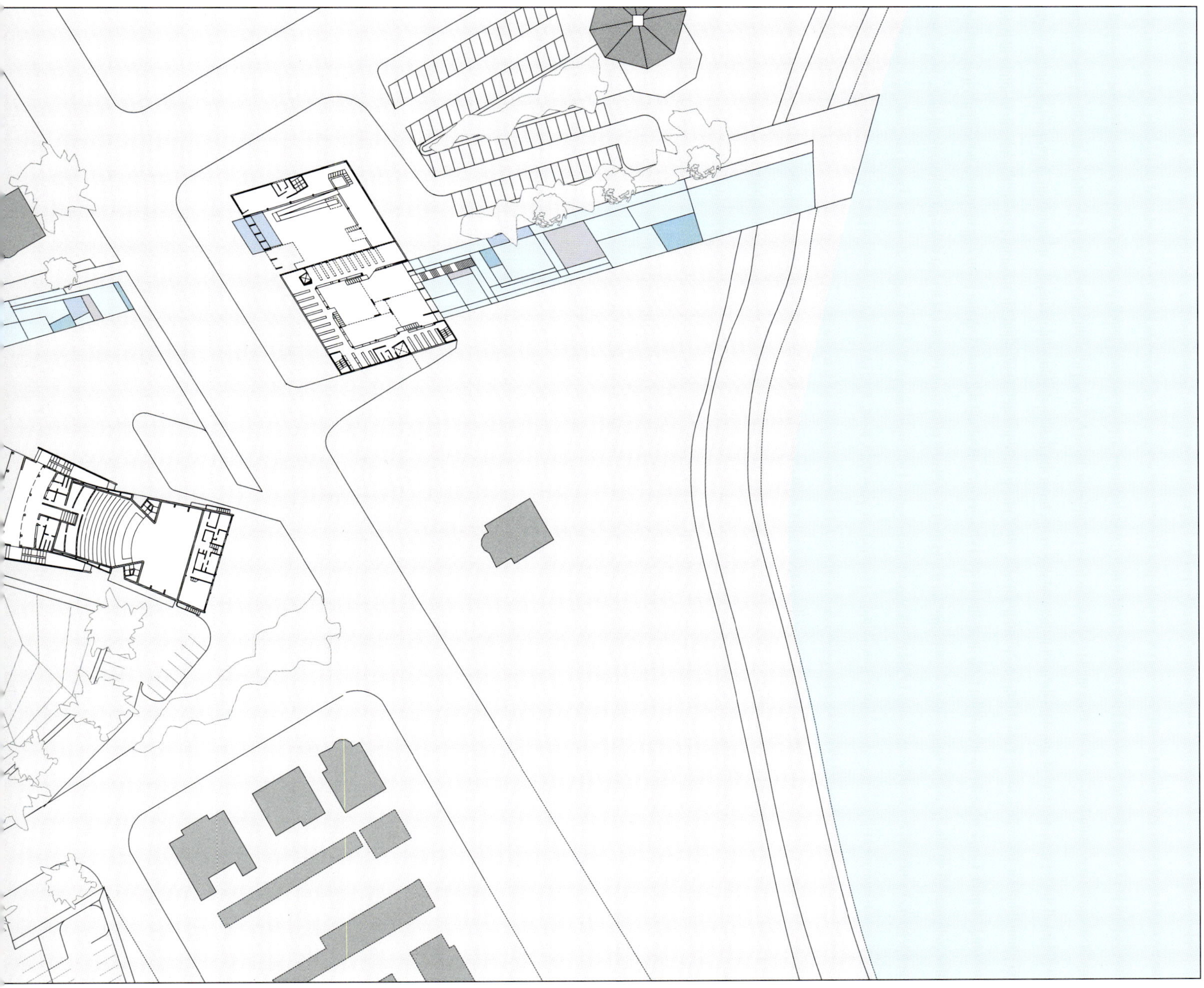

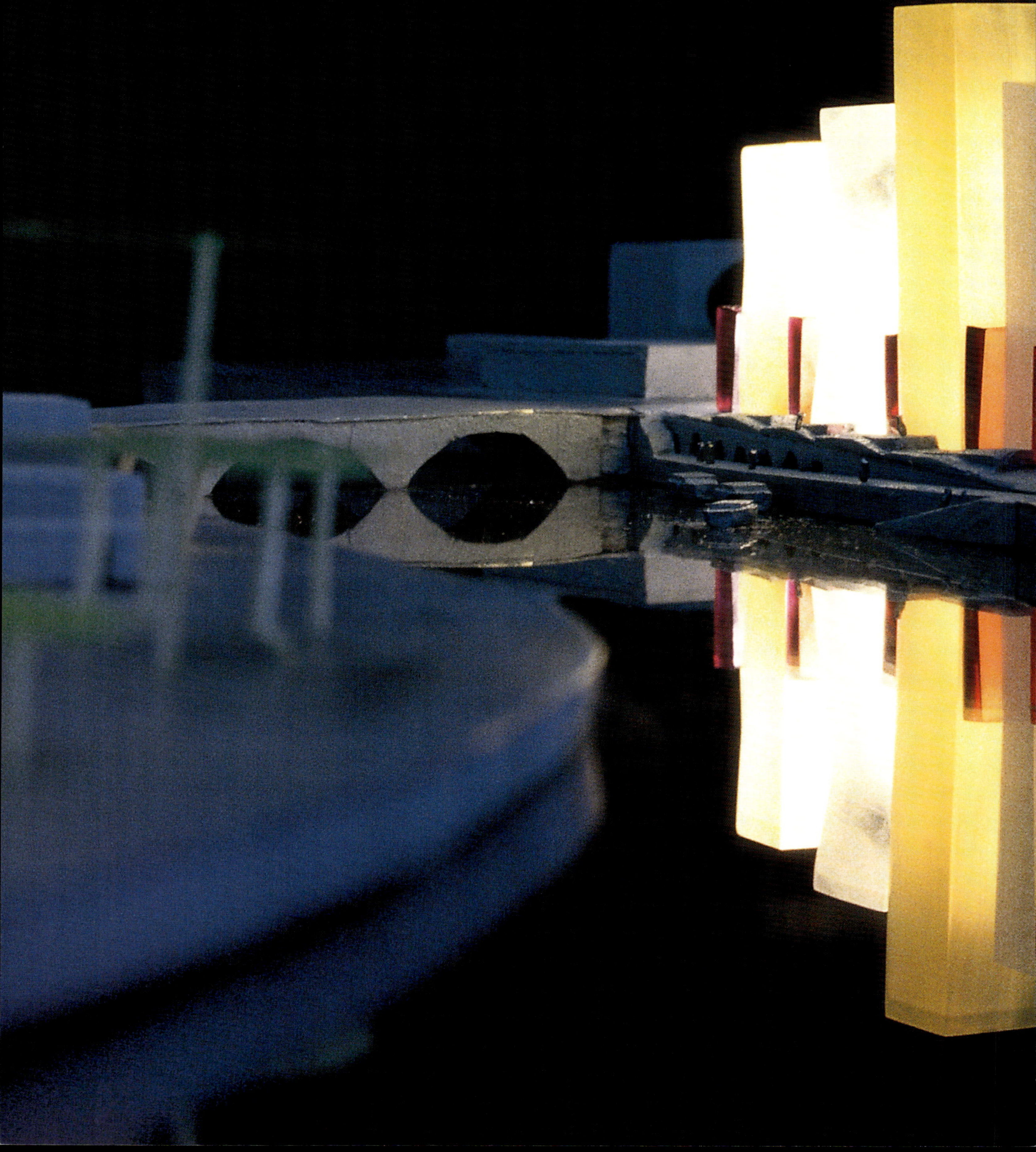

PROJECT NINE

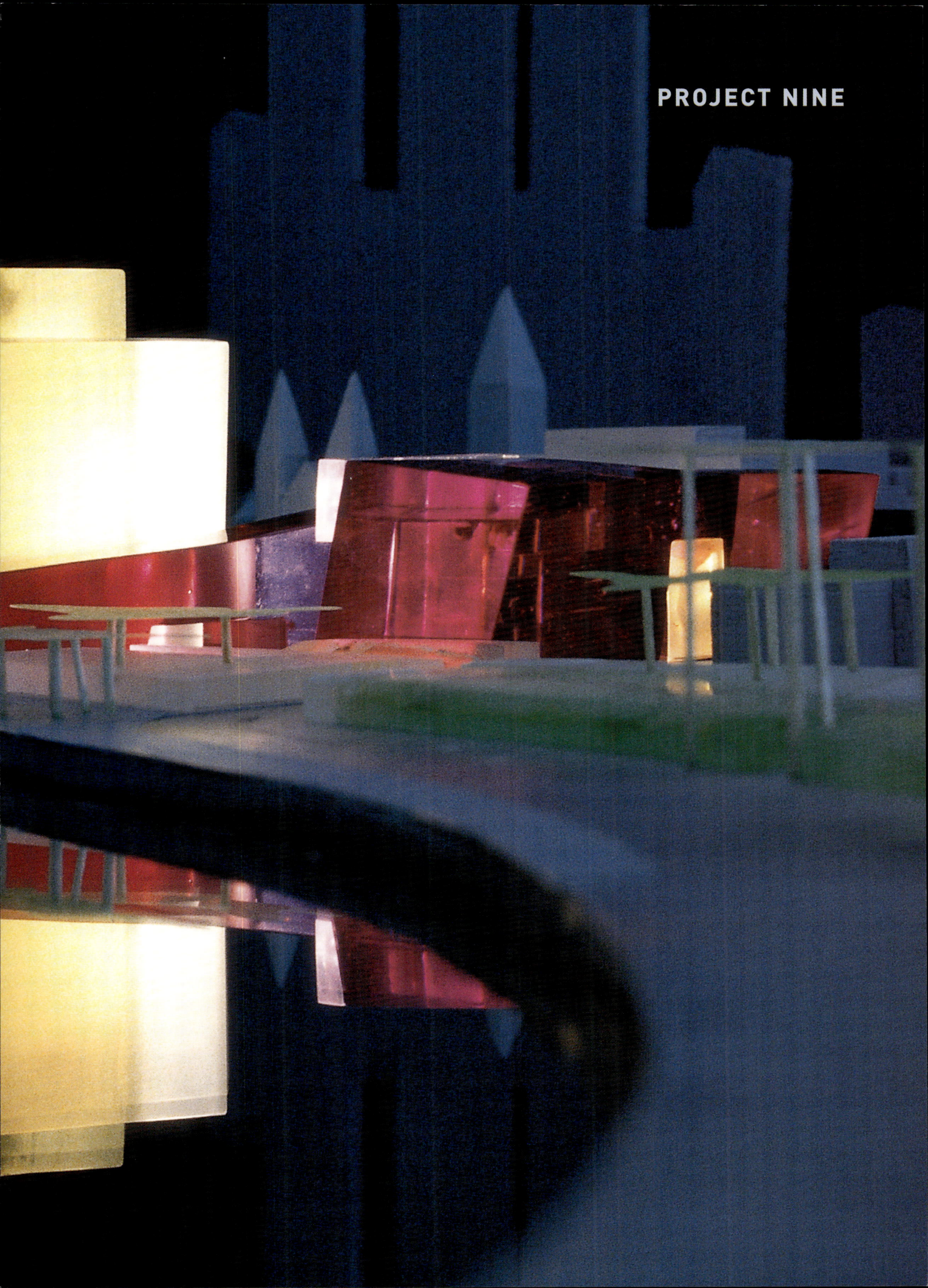

FEDERATION SQUARE MELBOURNE

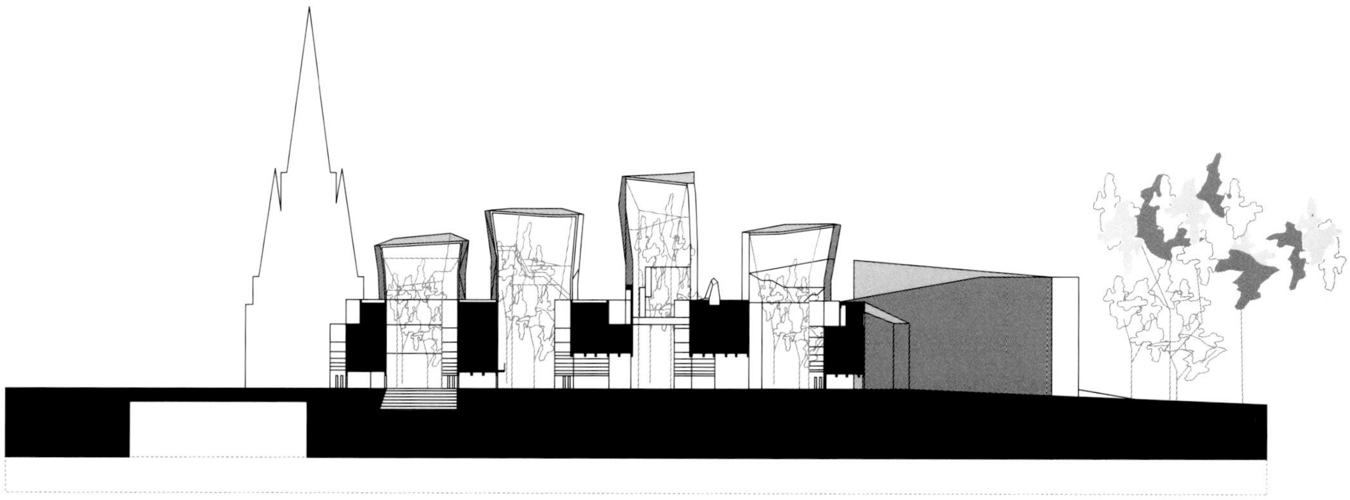

TWIST

The proposal straddles where figure and ground, city and park meet, twist and transform one another. Twisting through the city element, the park becomes a square: through the garden element, the city becomes a greenhouse. The intertwining is not simply about adjacency, but is an intense interchange of city and park, of the interior and exterior, of the possibility of collective belief and the pursuit of individual knowledge.

FIGURE AND GROUND

A square is a negative, an absence of building. It is formed by its surrounds, which shelter and animate, creating its civic constitution. The first move in the formation of the proposed square is a long building to the south along the river, sheltering the square from harsh winds. This building reconciles the need for shelter with the public impulse to move from the square towards the river. The scheme is based on the following propositions:

1. The ideal of the city, the collective, resides in the desire for density. This is expressed in the Helical building, where the individual programmatic requirements are fused together around a series of garden courts, where the spirit of the garden is held as if in amber.

The Helical Cinemedia and Performing Arts Centre, houses public presentation spaces and associated offices. Arranged in a loose helical structure, the mutual interaction of these spaces adds weight to each, binding them into a route of cultural immersion. The building serves as the pinwheel around which movement to and from the site pivots. Arrivals occur on all its faces: the multiple entrances underscore its essentially public quality, encouraging movement through and about its voids and clusters.

2. The ideal of the garden is the desire for solitude, a foreground holding a captive horizon. Buildings are set in the garden as places of momentary repose under the sky. This is the opportunity of the Wintergarden, where the contemplation of culture takes place within a heady profusion of greenery.

The Wintergarden holds a large public space overlooking the Yarra. Above are the galleries, connected by walkways. Still higher are the four soaring glass lanterns, accommodating the highest tree canopies and the topmost resting platforms. A double skin of glass allows structure and services to hide mysteriously, technology simply doing its work without advertising its aesthetic. The possibility of opening up large portions of each skin with doors or a sliding bifold roof enables the effortless transformation of an interior to an exterior.

PROJECT TEAM Neil Durbach, Camilla Block, Catherine Lassen, Lisa Le Van, Harry Margalit 1997

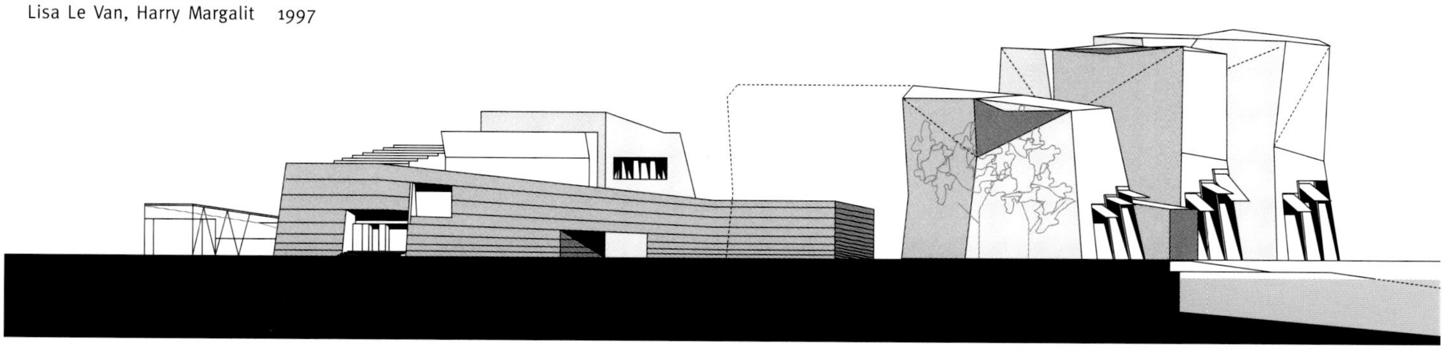

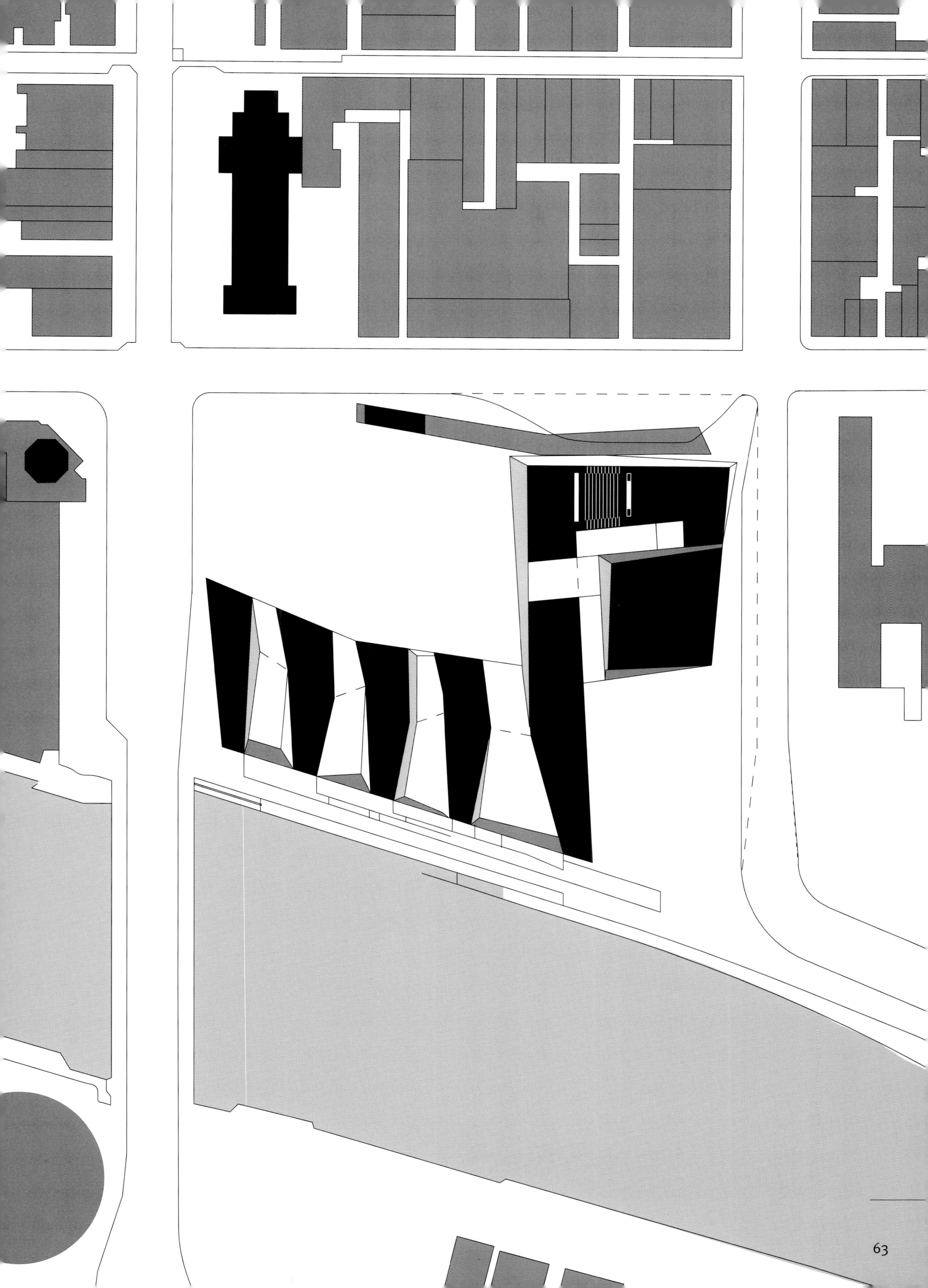

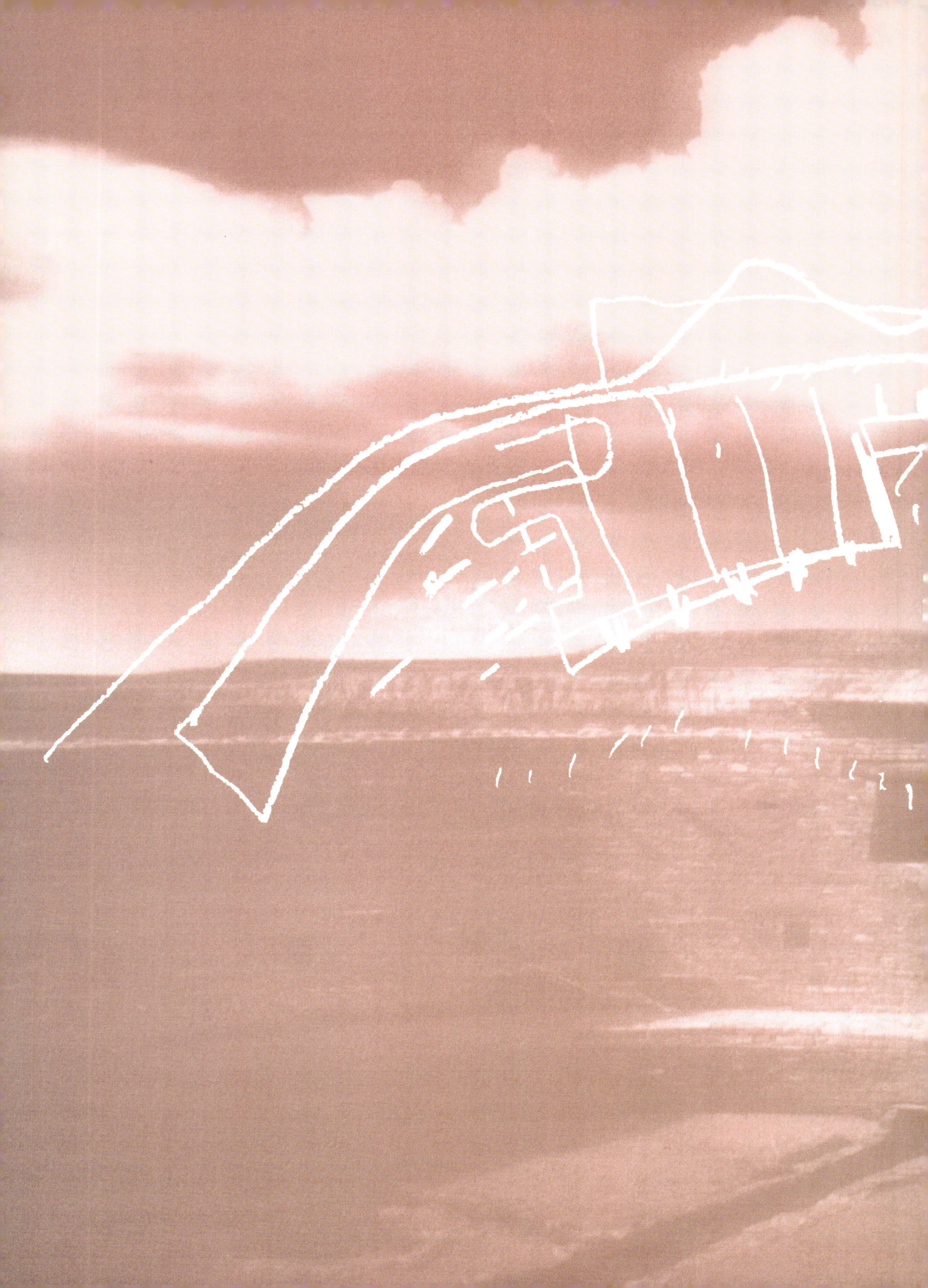

PROJECT TEN

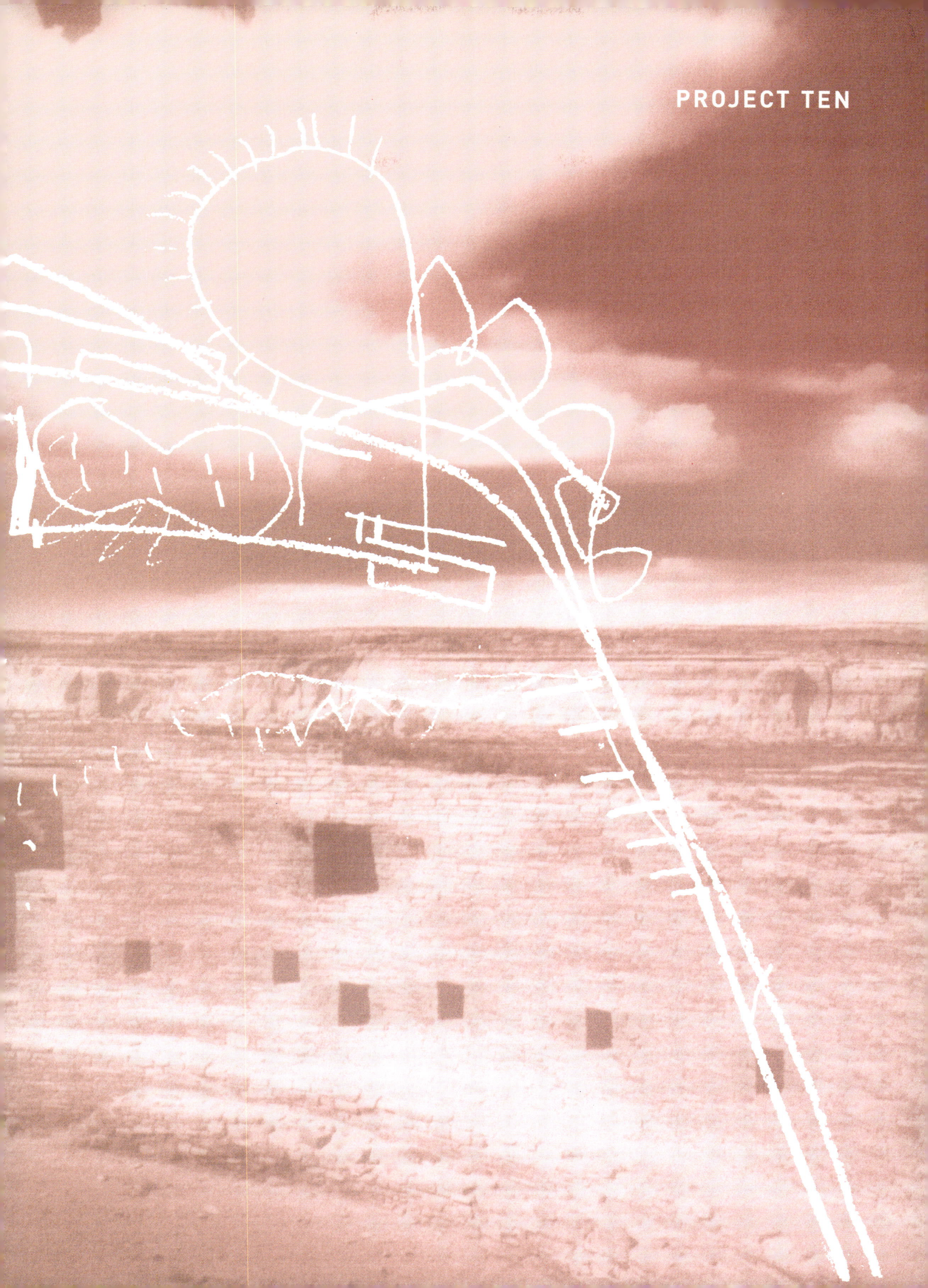

HUNTERS HILL HOUSE HUNTERS HILL, SYDNEY

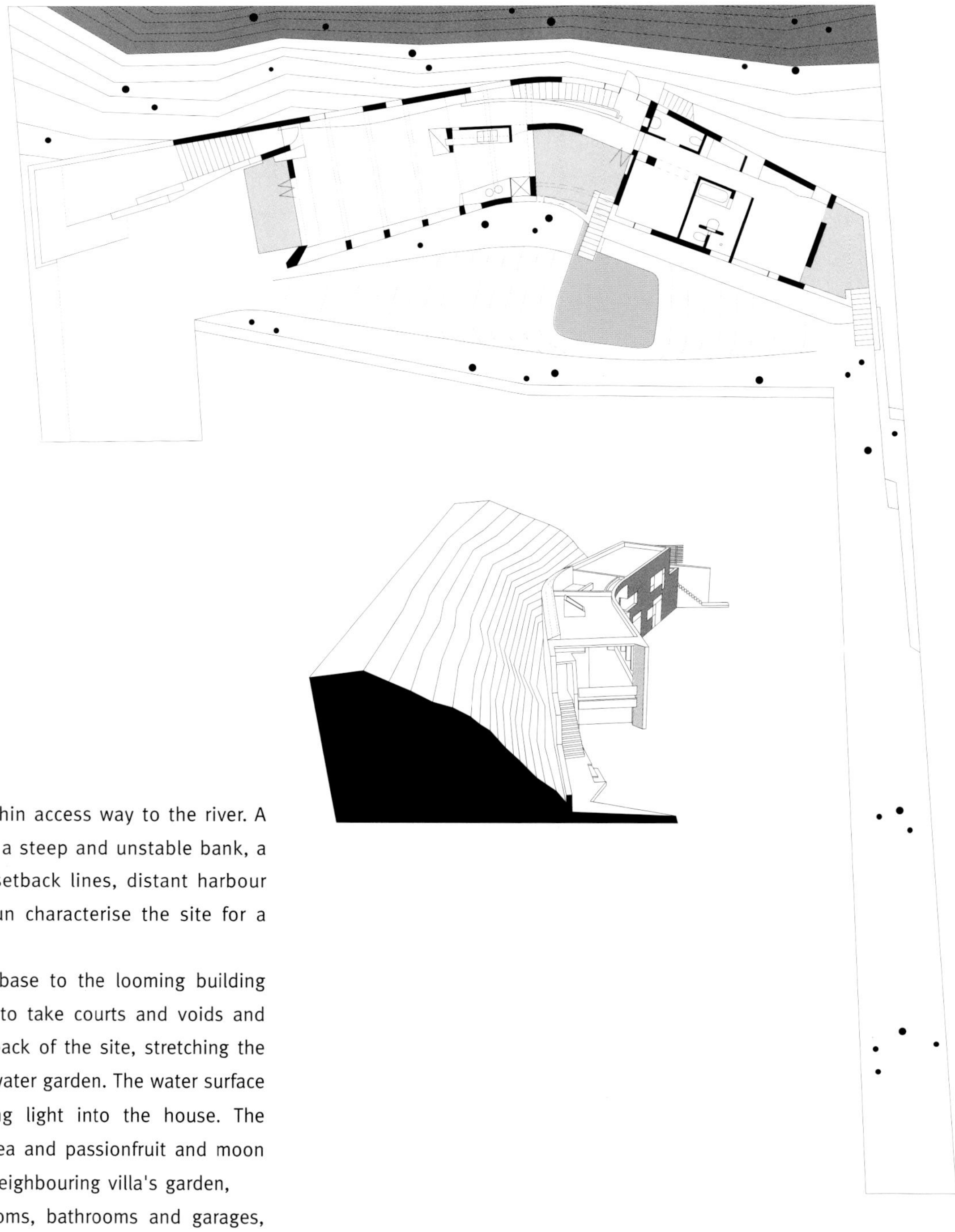

A south facing site with a pencil-thin access way to the river. A Late Federation villa to the north, a steep and unstable bank, a mesh of controlling heights and setback lines, distant harbour views and limited moments of sun characterise the site for a family home.

The house is conceived as a base to the looming building behind, a dam wall hollowed out to take courts and voids and slices of light. It arcs around the back of the site, stretching the view and clearing space for a vast water garden. The water surface acts as a rippled mirror, reflecting light into the house. The house's roof garden of bougainvillea and passionfruit and moon flowers is perfectly level with the neighbouring villa's garden,

The lower floor houses bedrooms, bathrooms and garages, and serves as a miniature walled space to the living spaces, main bedroom and study above. Three terraces punctuate the arc. One to the west with a view of Gladesville Bridge, another private with a narrow stair to the river, and a third central cut that connects the living spaces to a paved island in the water garden, capturing a slice of northern sun in winter.

The intention is to make the house feel weighted and grounded, both at ease and respectful.

PROJECT TEAM Neil Durbach, Camilla Block, Lisa Le Van, Joseph Grech. Date of completion 1998

SOUTHERN FACADE, WATER GARDEN AND PAVED ISLAND

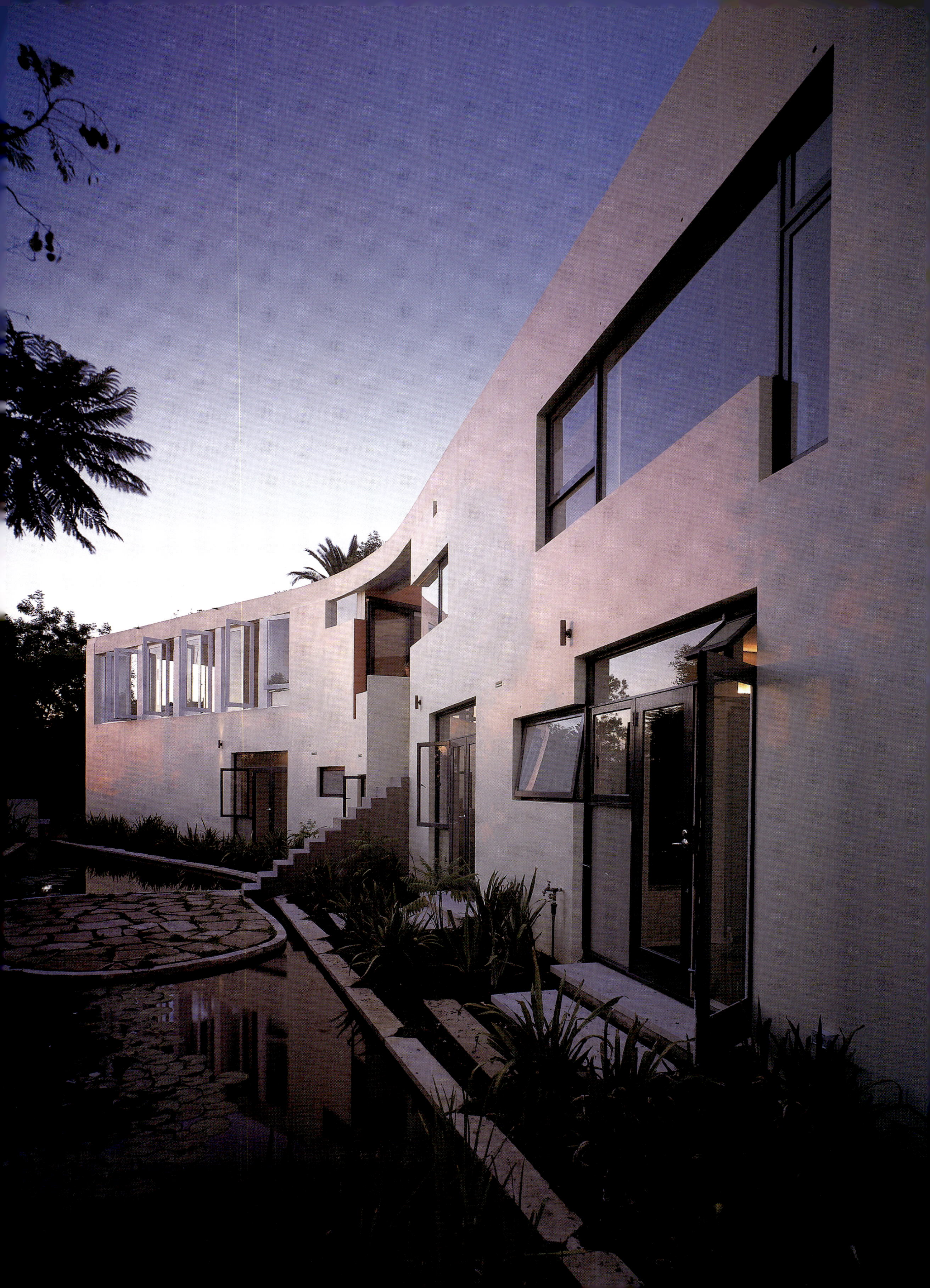

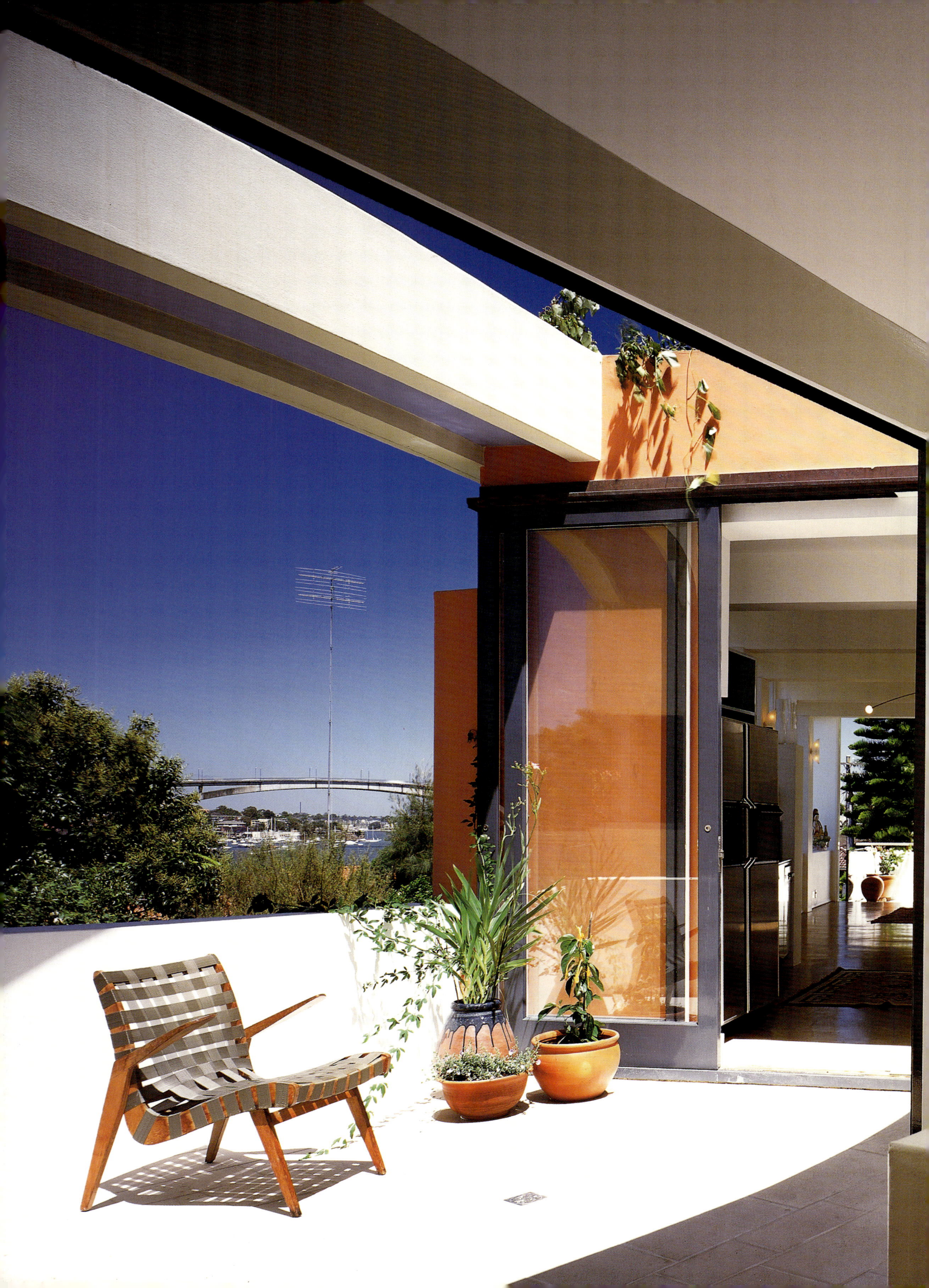

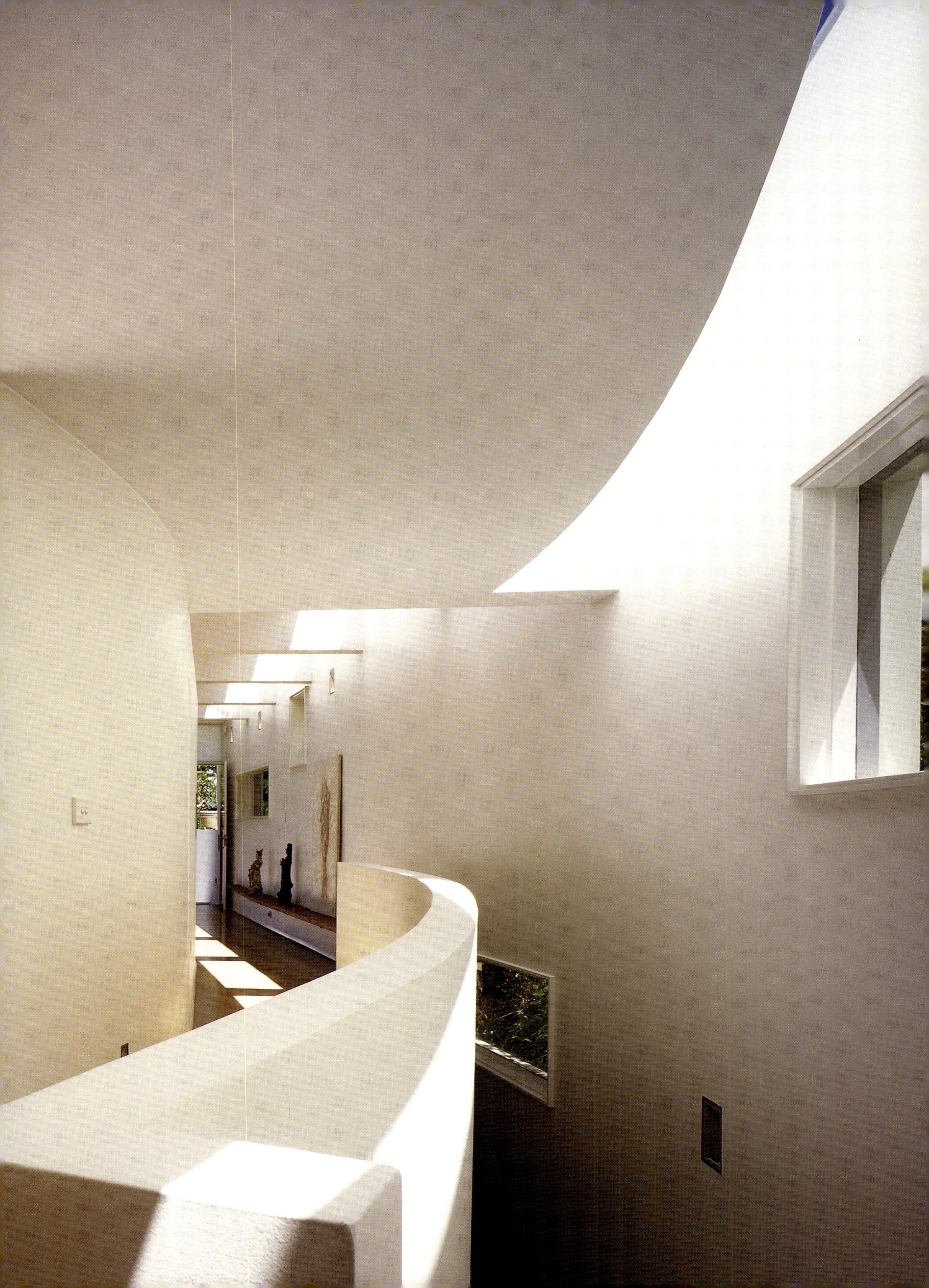

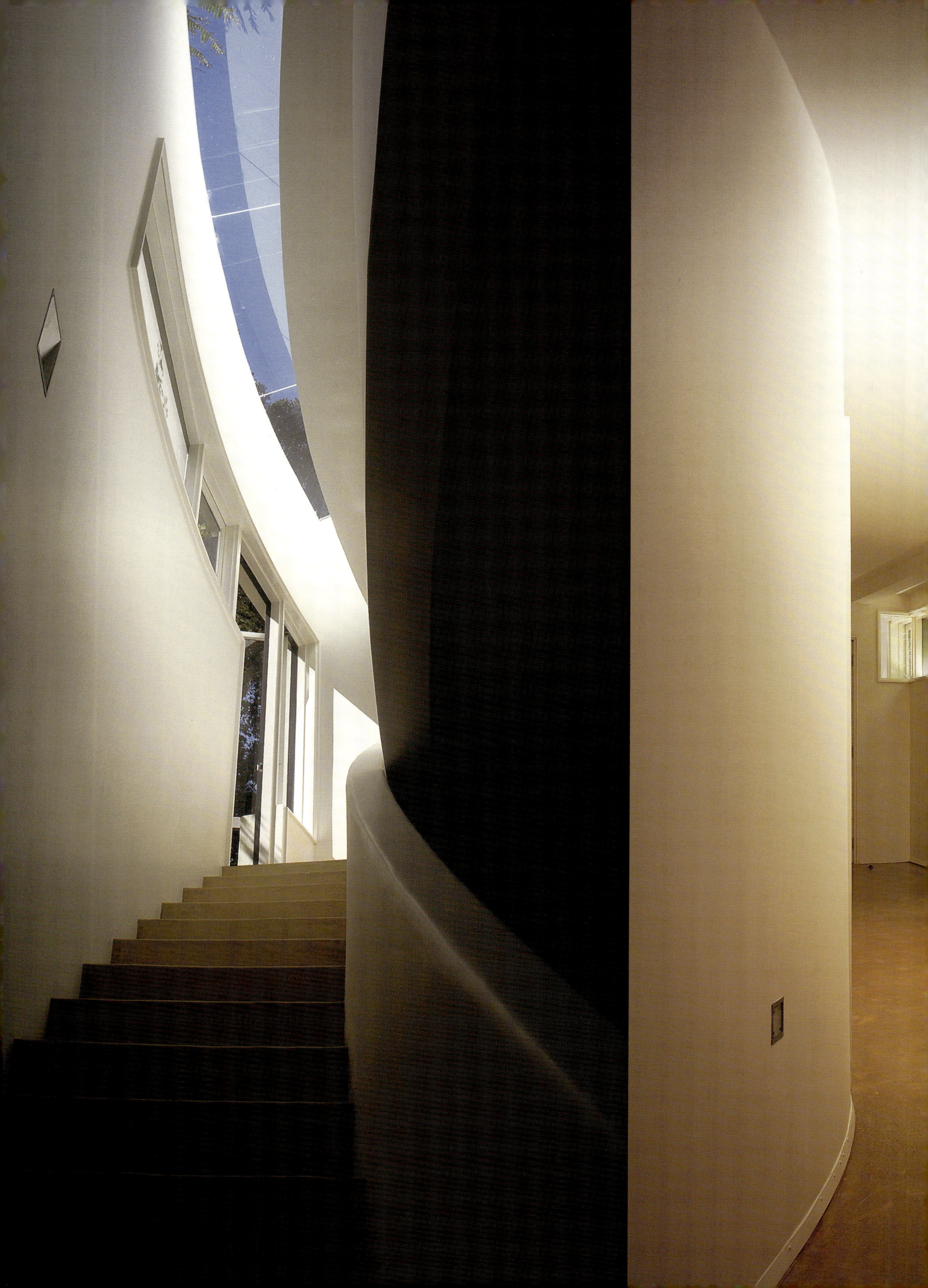

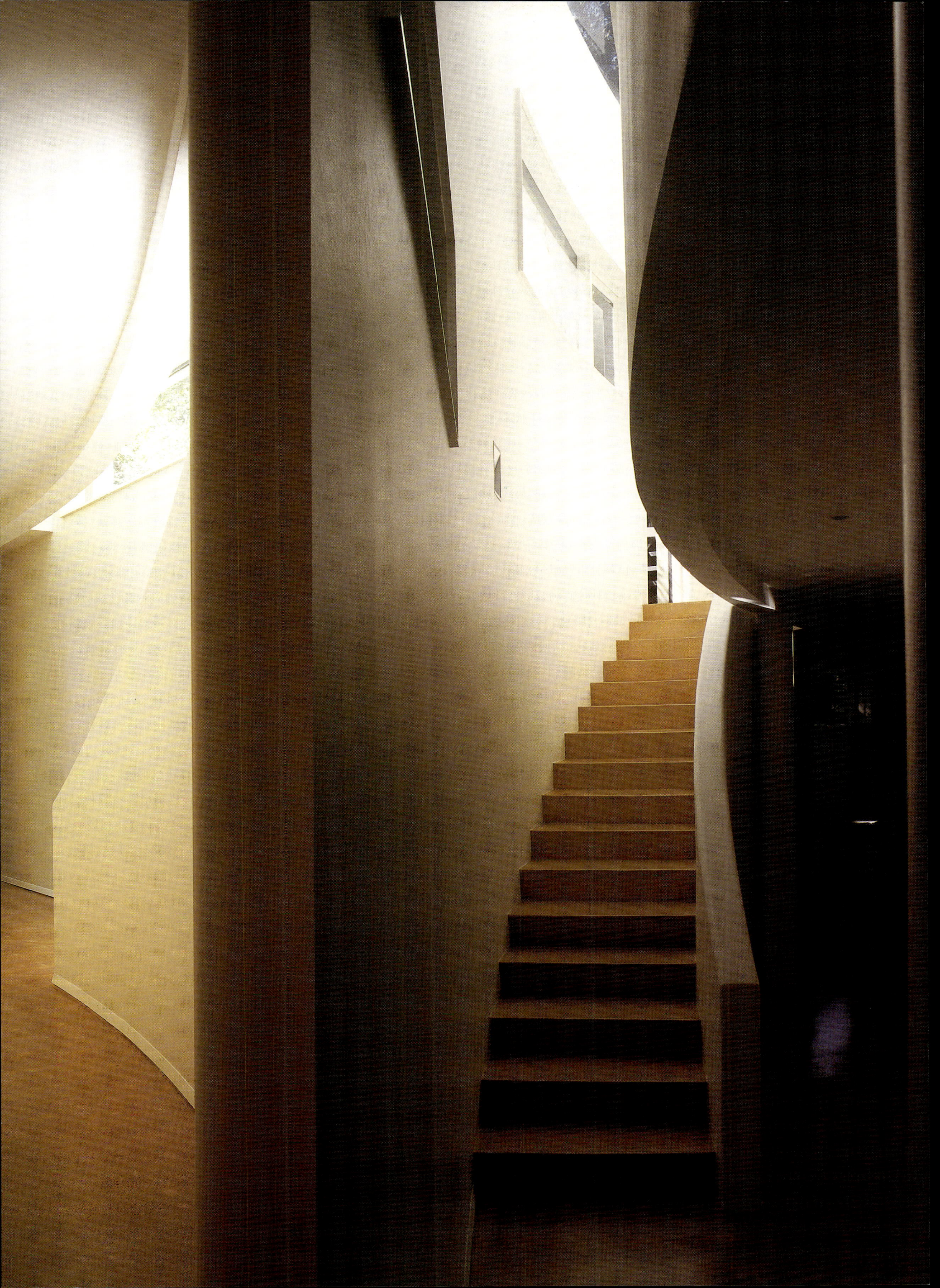

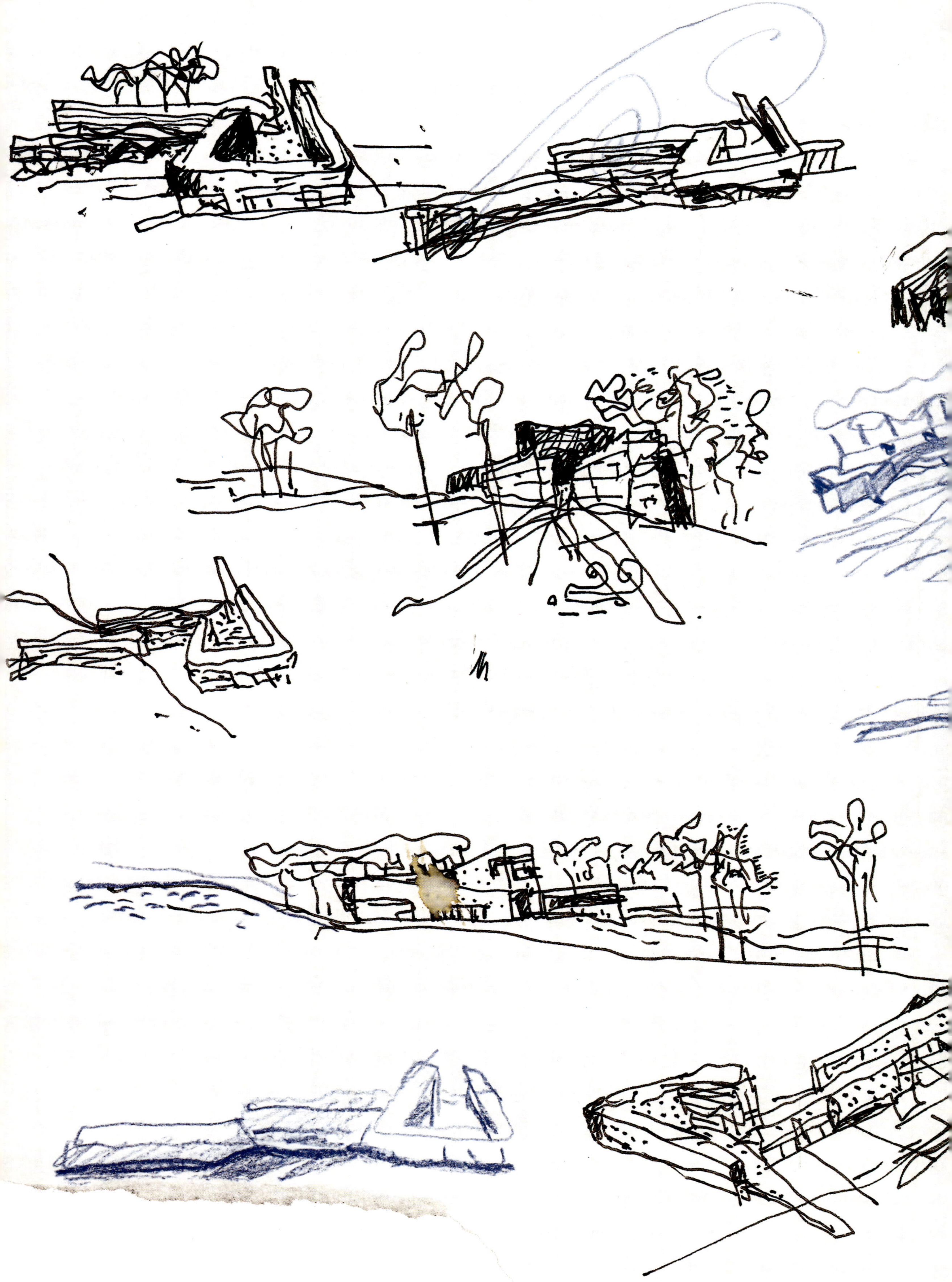

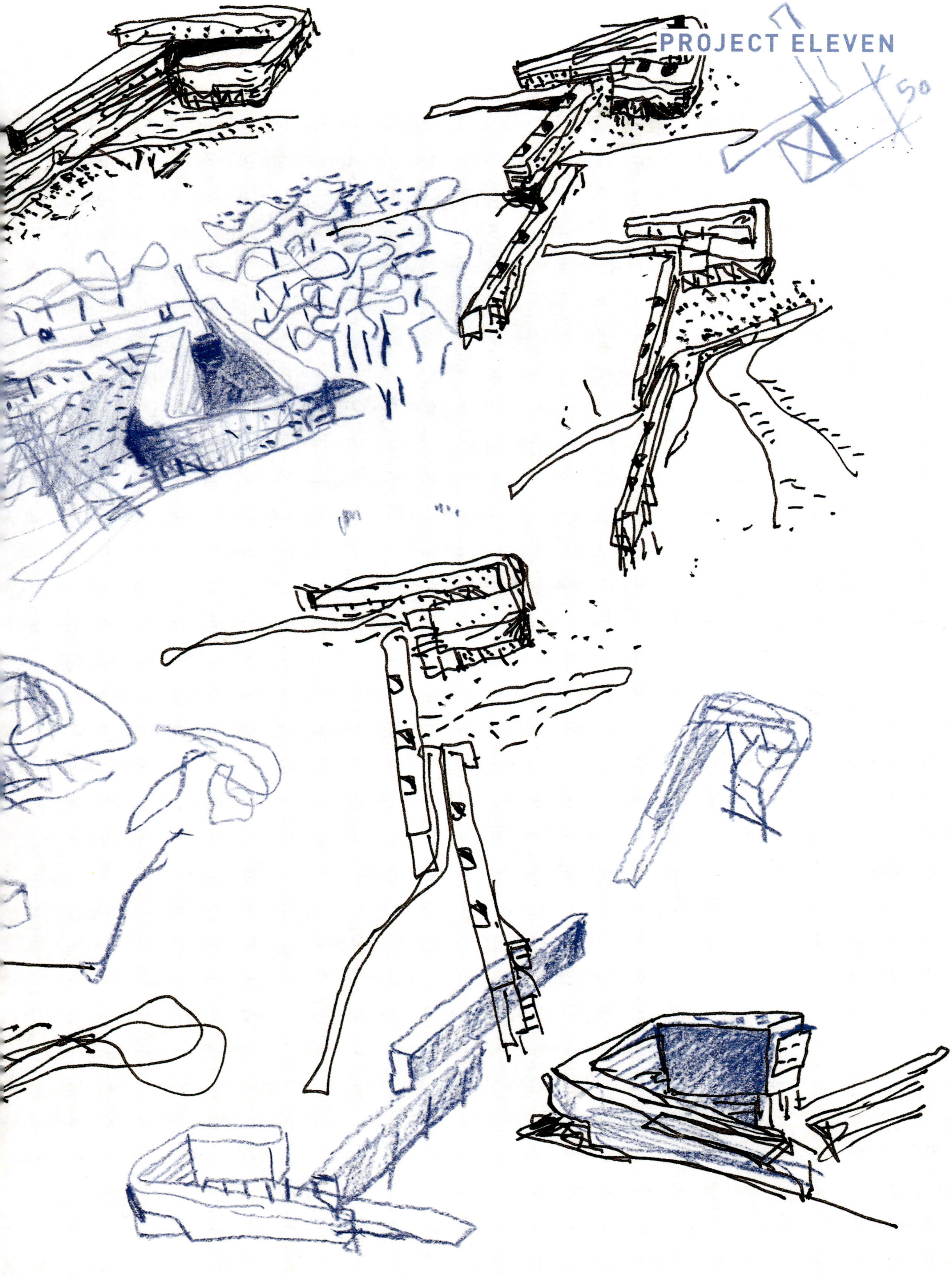
50

STUDENT HOUSING BUNDOORA CAMPUS RMIT TECHNOLOGY ESTATE

The brief called for staged student housing for the Royal Melbourne Institute of Technology at their campus on the northern edge of Melbourne. The scheme is an overlay which clarifies existing attributes of the site and its masterplan. The site will be structured by the establishment of magnificent gardens as a first and necessary move. The gardens are structured by the gathering, cleansing and redistribution of water. The housing is structured by an understanding of shared amenity and living with nature. The rooms are structured by an elastic relationship between the individual and the group. Stage One of the housing is a collegiate courtyard cracked open to increase exposure to the North and expand the variety of configurations. Miniature courtyard types are repeated into lengths to enable various and complex reconfigurations. Within this model are gradients of private and communal spaces, ranging from shared stairwells and tightly jacketed garden courts to wide corridors for lounging and conversation - an array of spaces suited to a plenitude of moments.

PROJECT TEAM Neil Durbach, Camilla Block, David Jaggers, Chris Metz, Lisa Le Van, Nancy Lim 1999

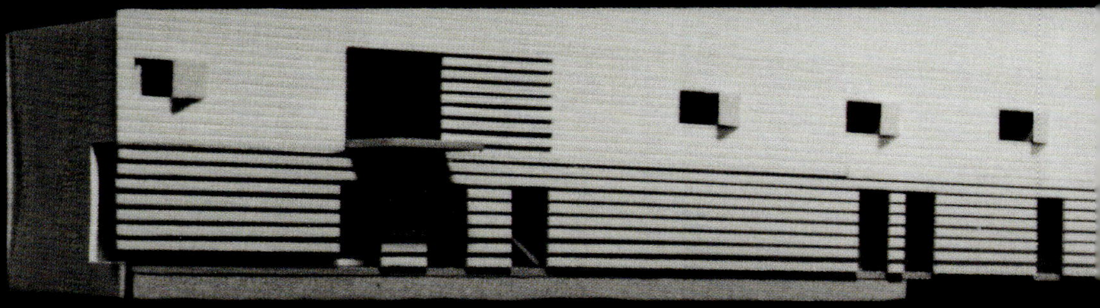

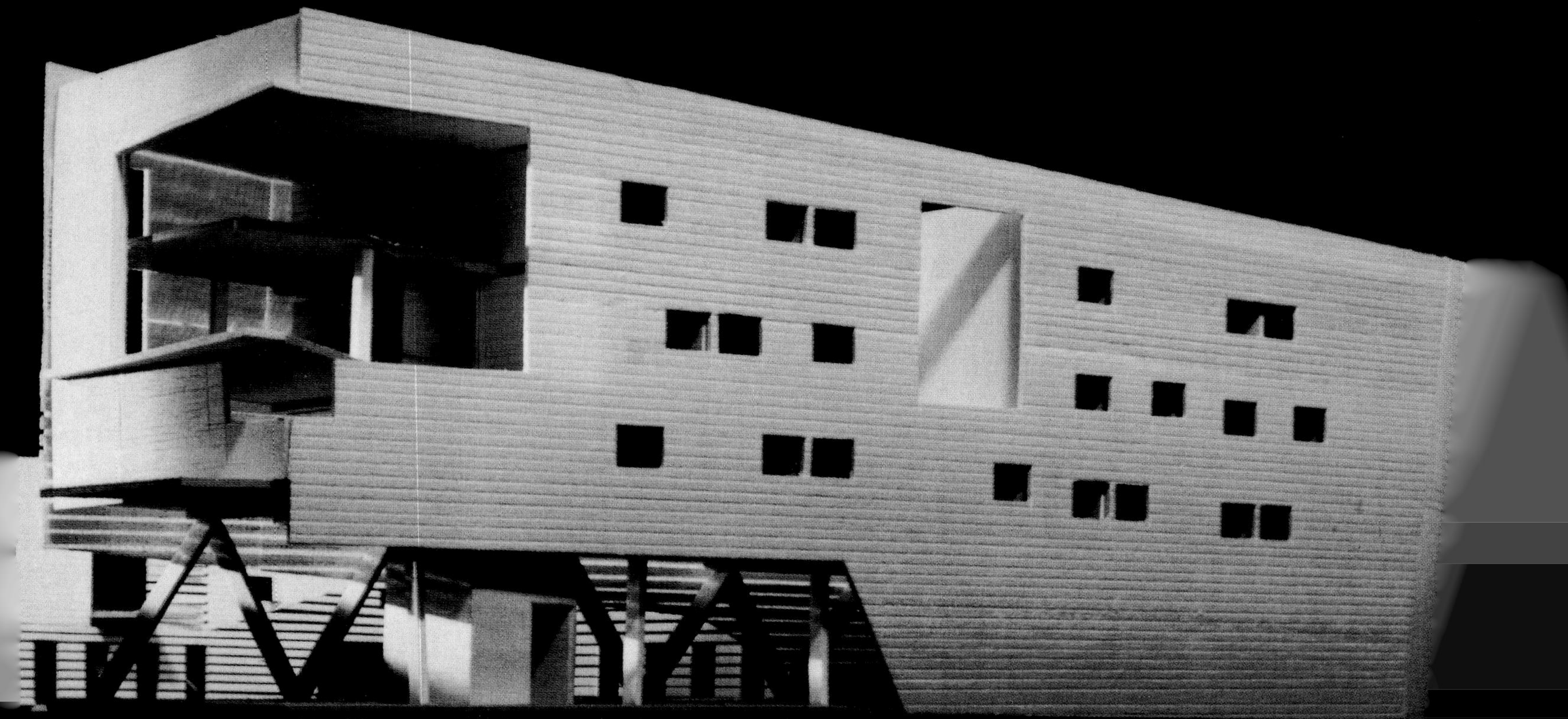

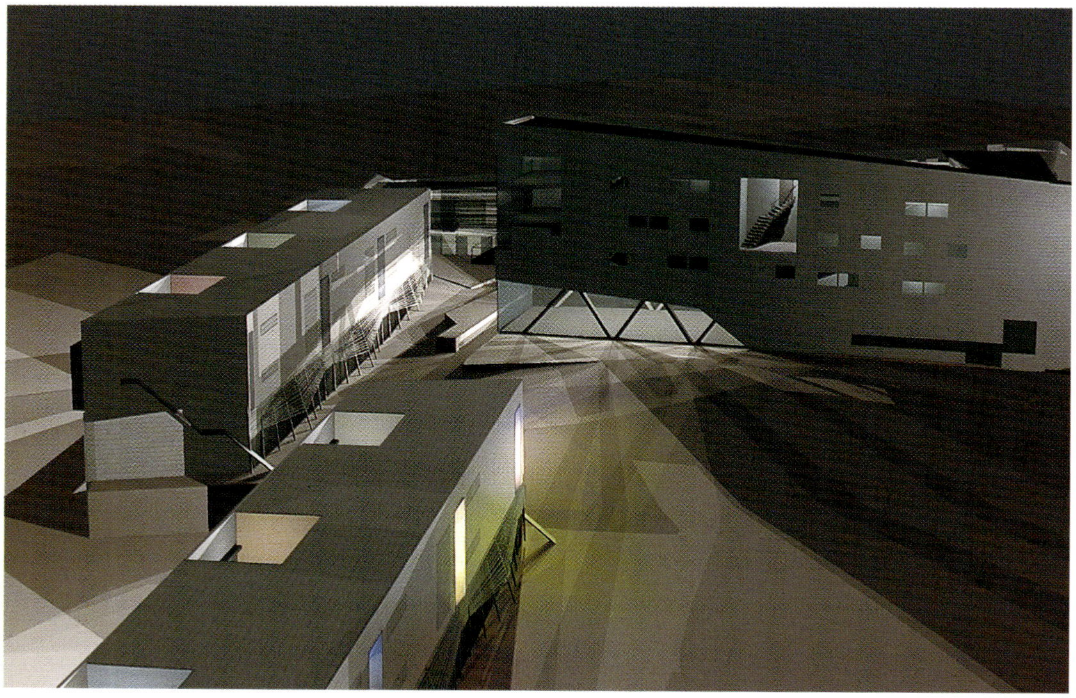

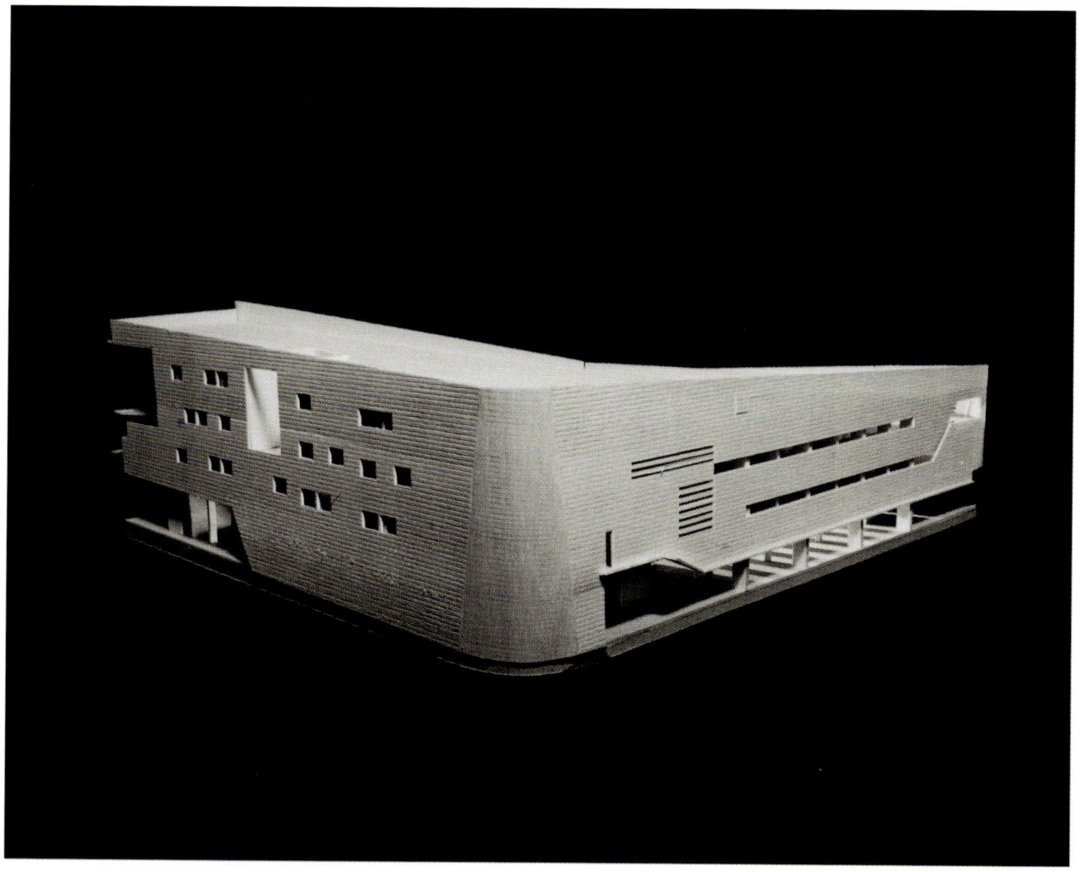

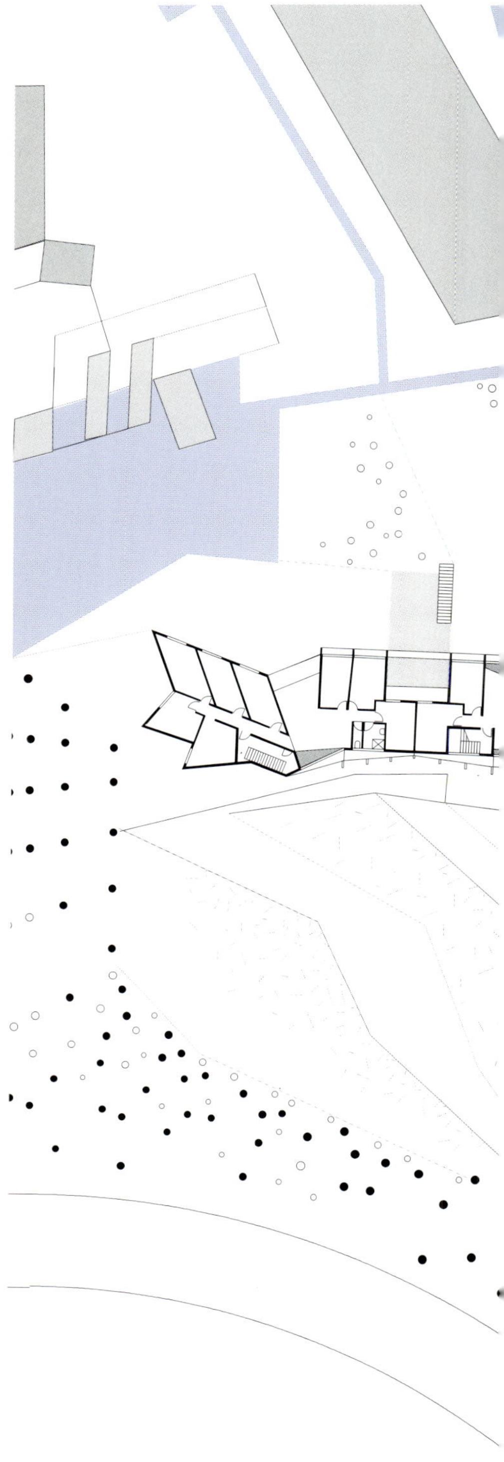

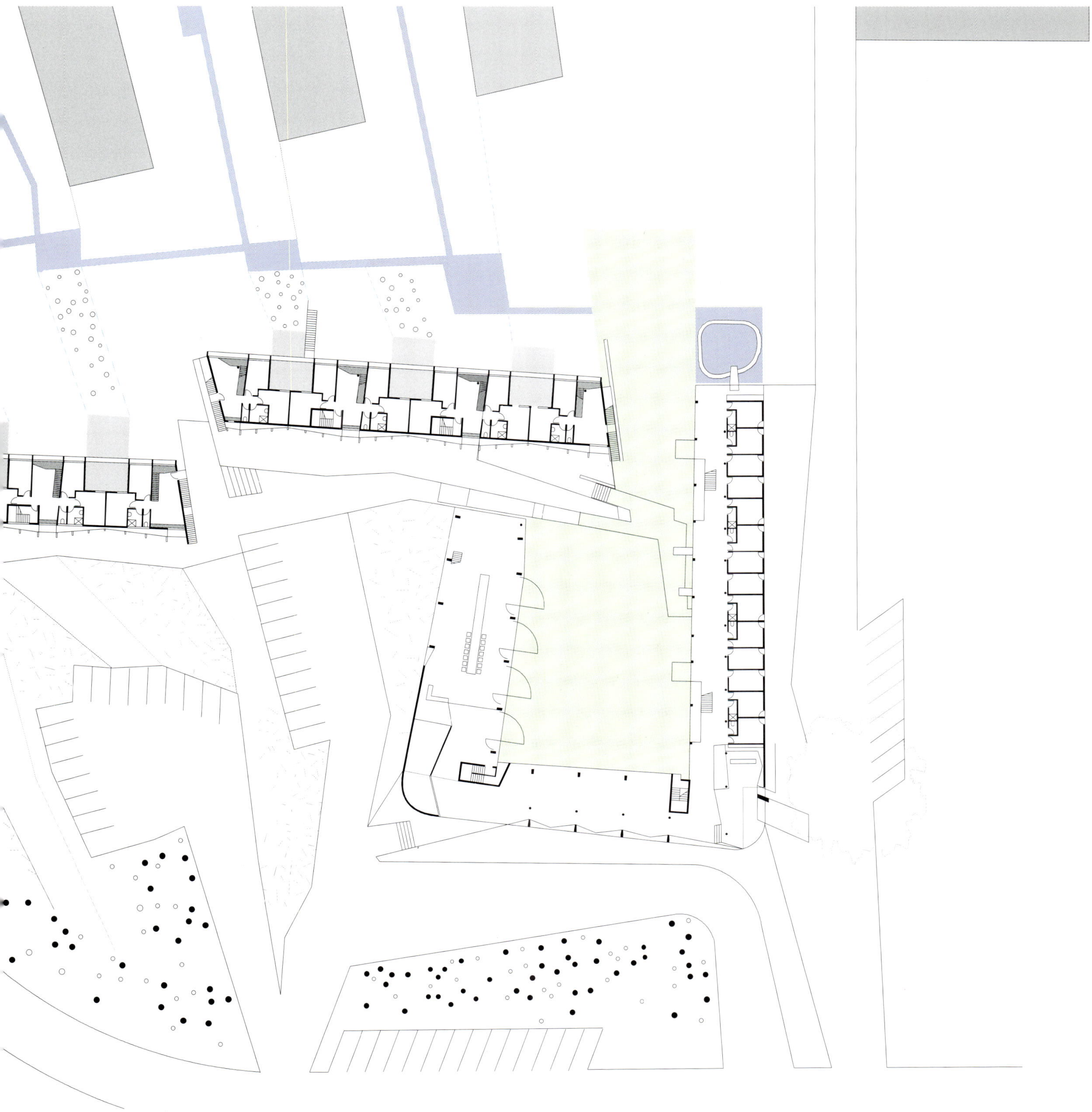

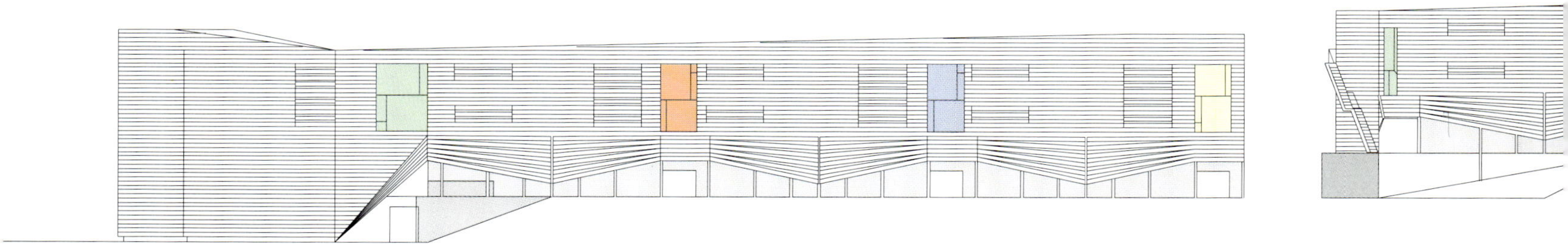

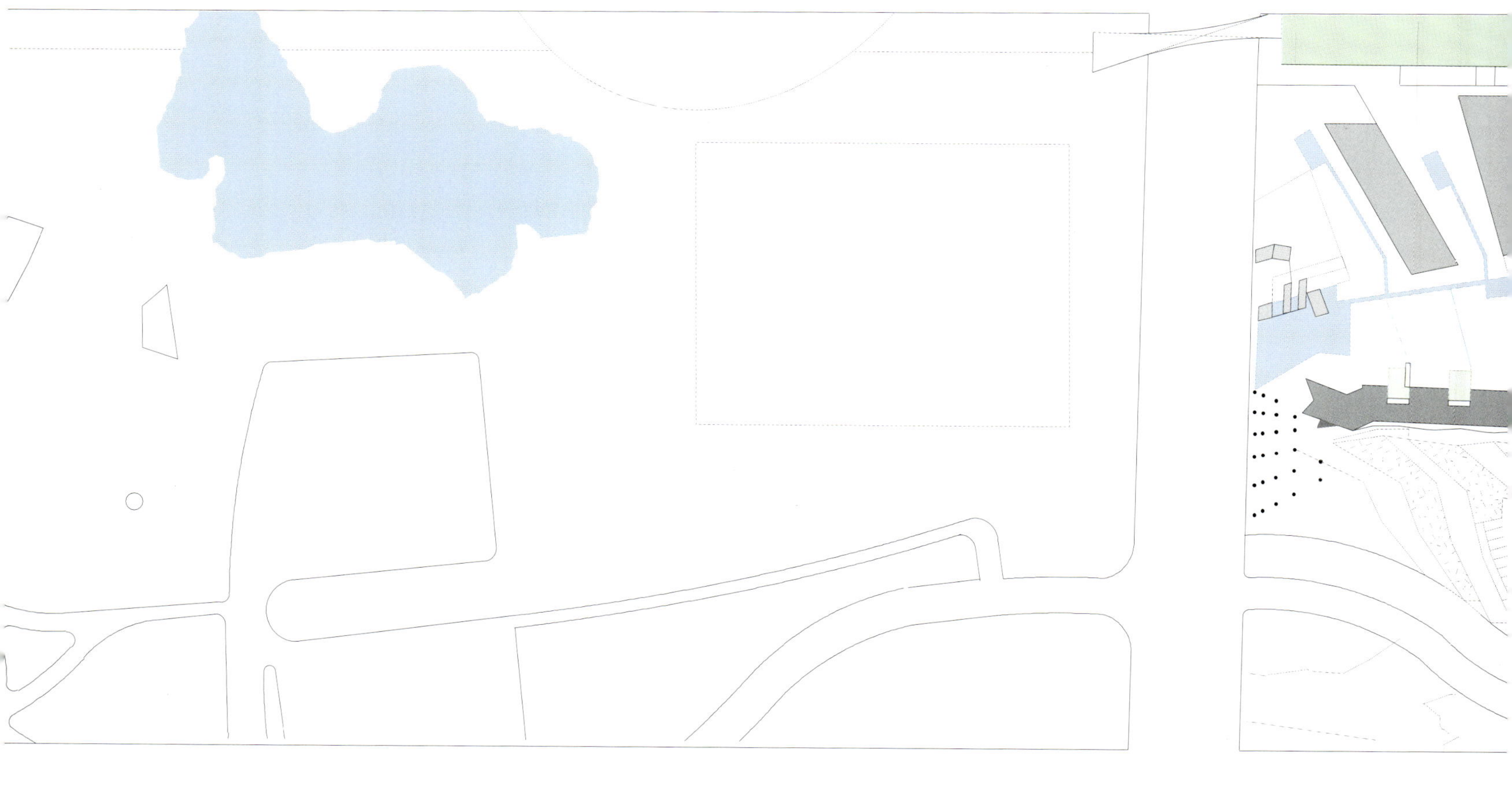

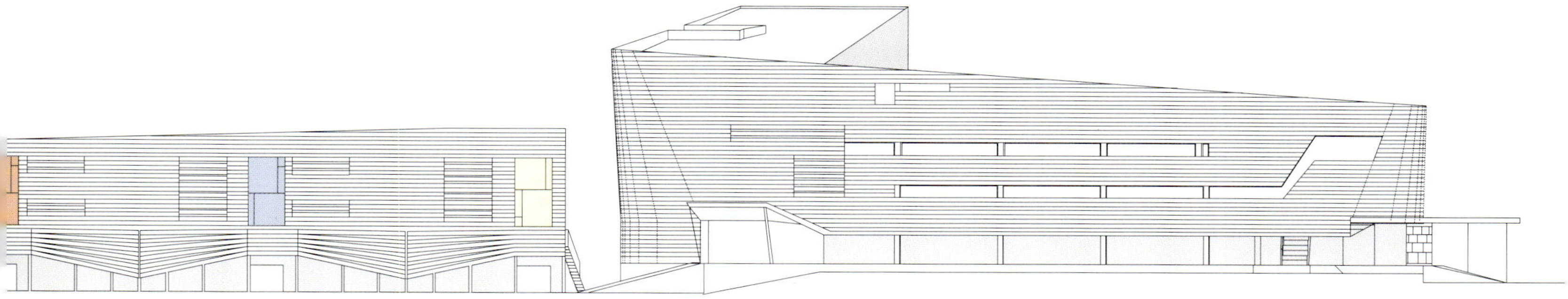

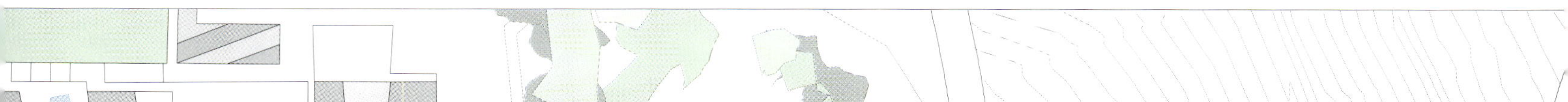

VISTA STREET GREENWICH

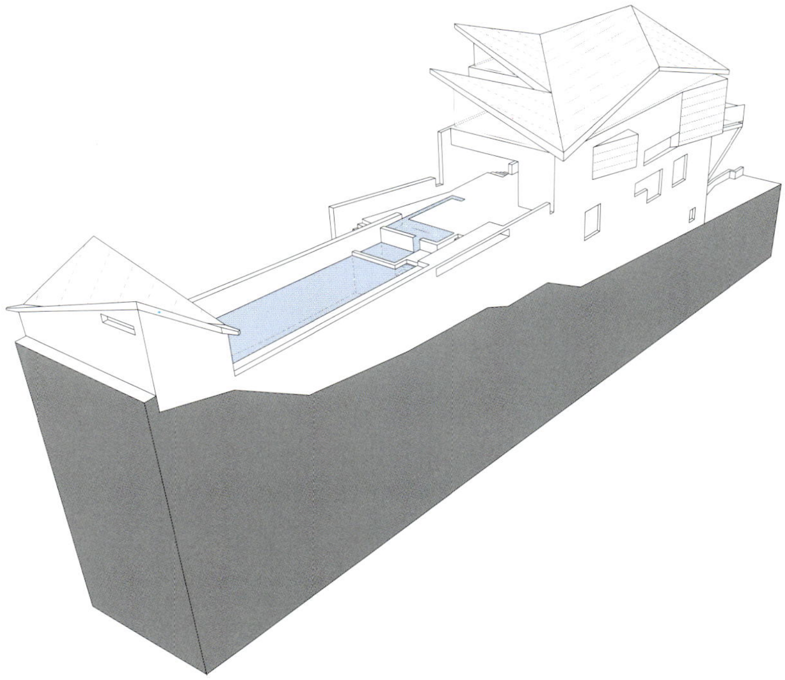

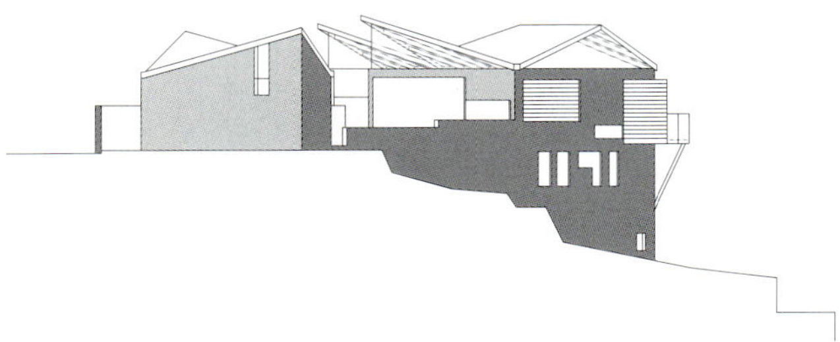

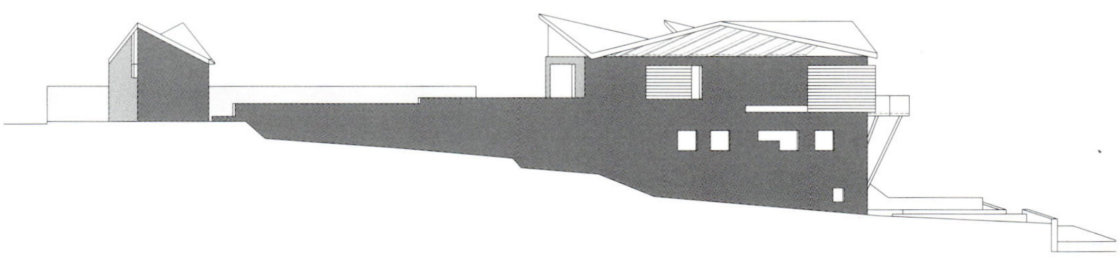

A suburban house on a raked block with views to the Harbour and the Bridge. The site falls towards the view, away from the sun. Held between the view and a north facing garden, the section works its way from the road to the pool pavilion. With bedrooms on the lower level, the living area makes an easy transition from the view to the pool and adjoining garden. Its roof, folded rather than pitched, unsettles an orthogonal plan and grabs segments of sun, view and sky. The garden, scented and enclosed, cuts deep into the plan and section.

PROJECT TEAM Neil Durbach, Joe Grech, Camilla Block, Lisa Le Van

Date of completion October 1999

PROJECT TWELVE

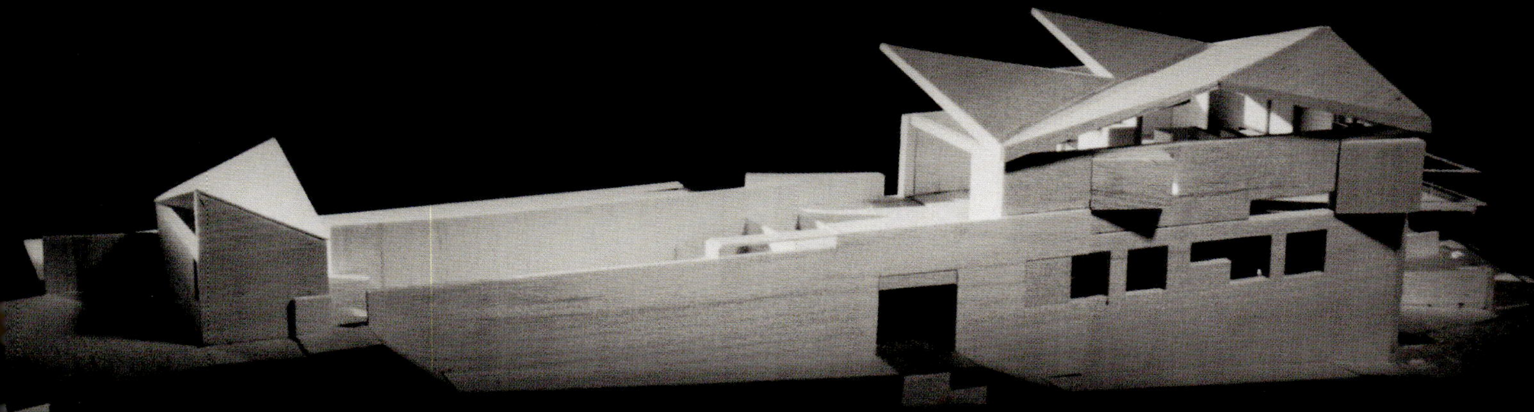

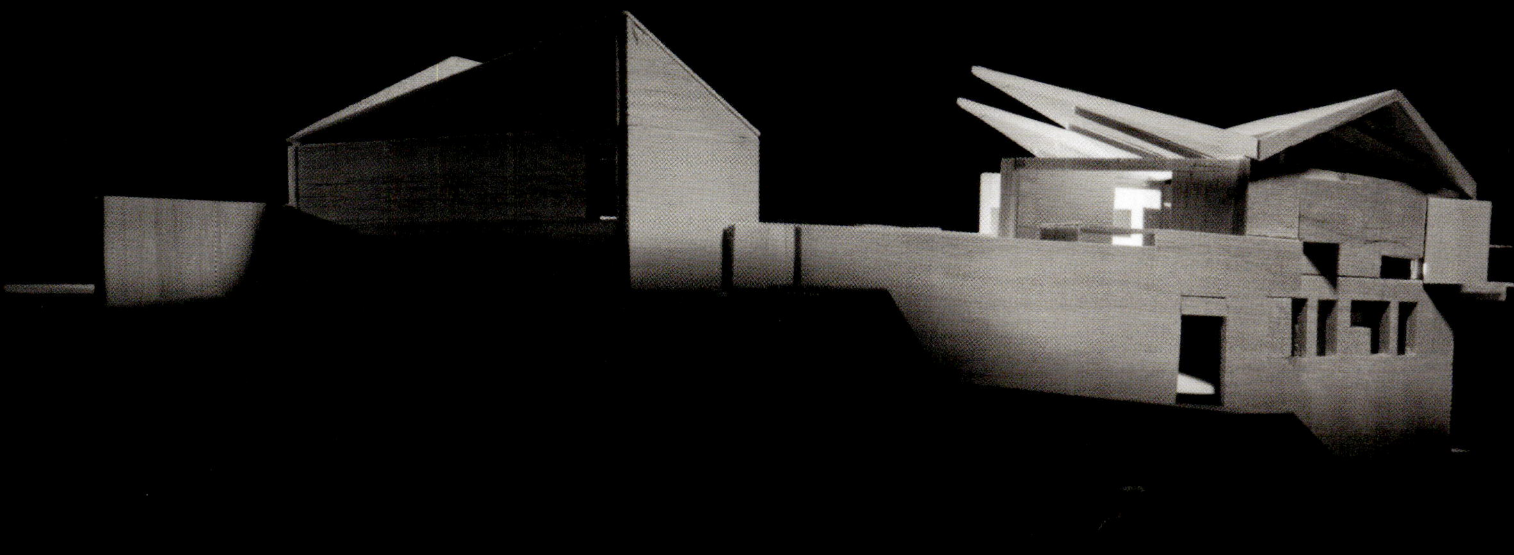

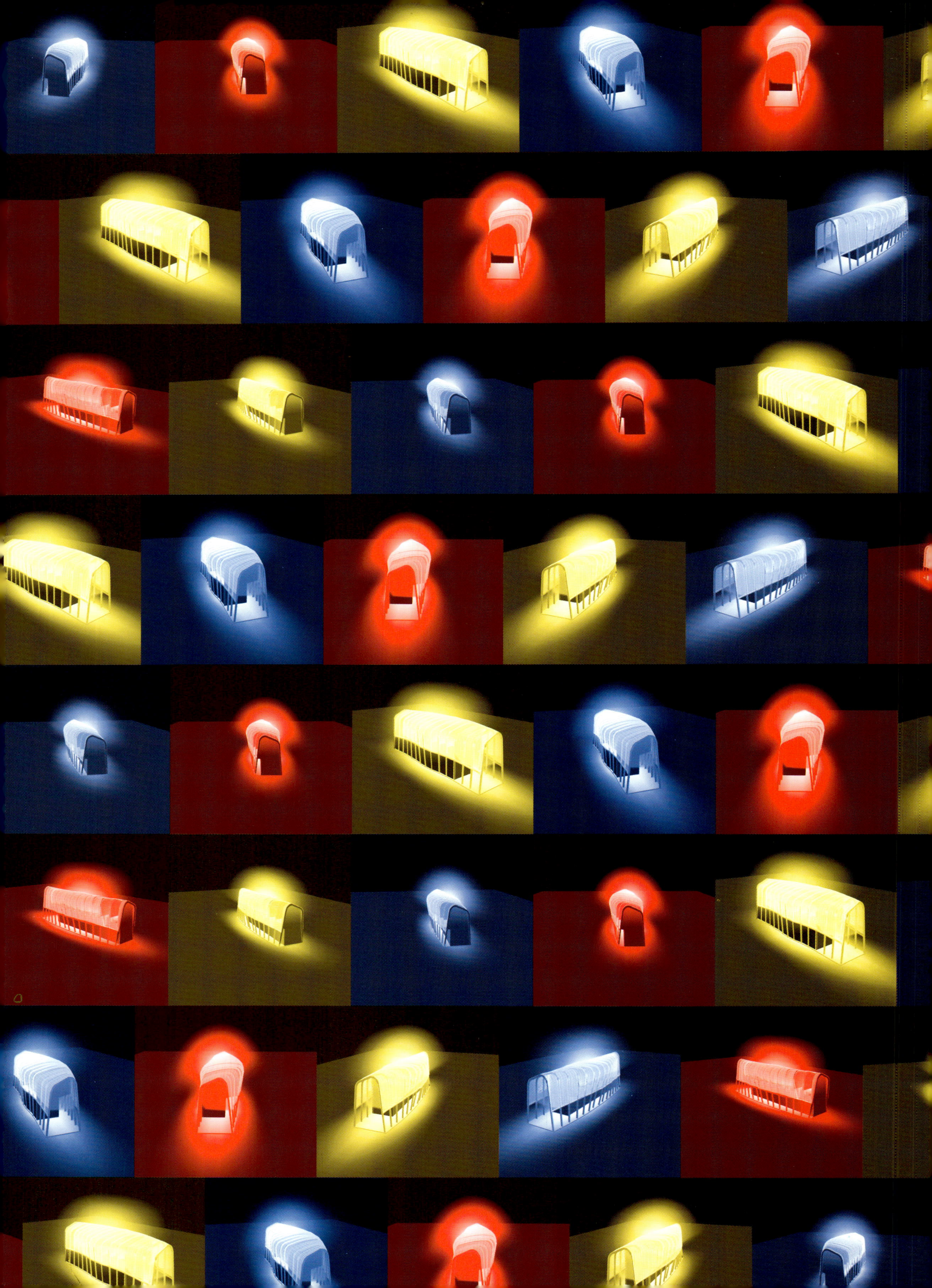

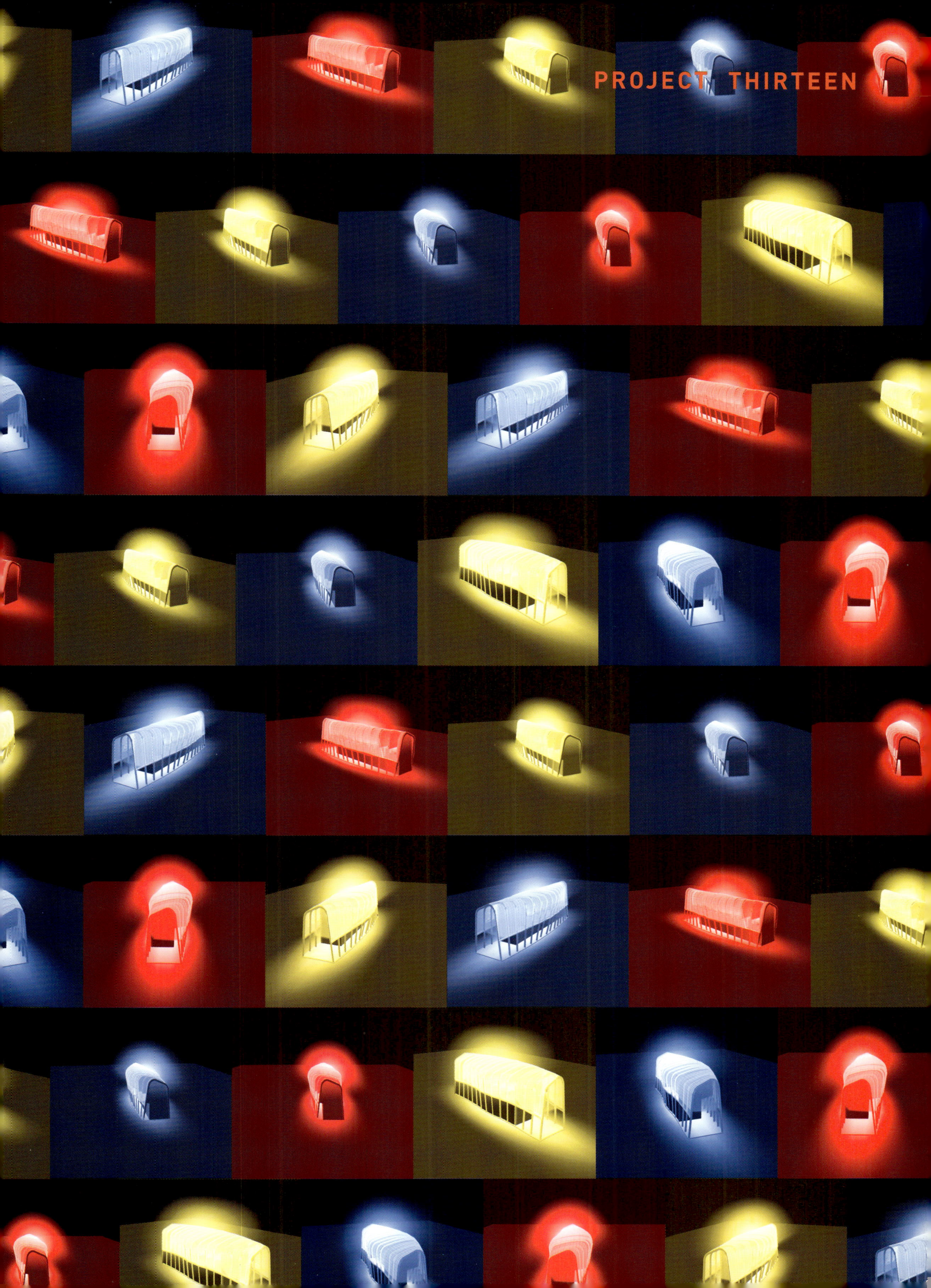
PROJECT THIRTEEN

AMENITIES BUILDINGS, SYDNEY OLYMPIC SITE HOMEBUSH BAY, SYDNEY

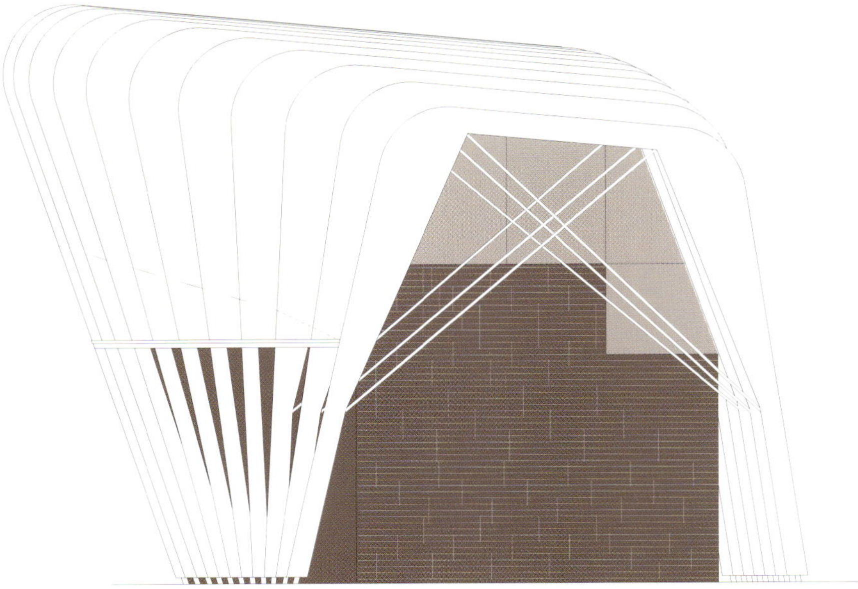

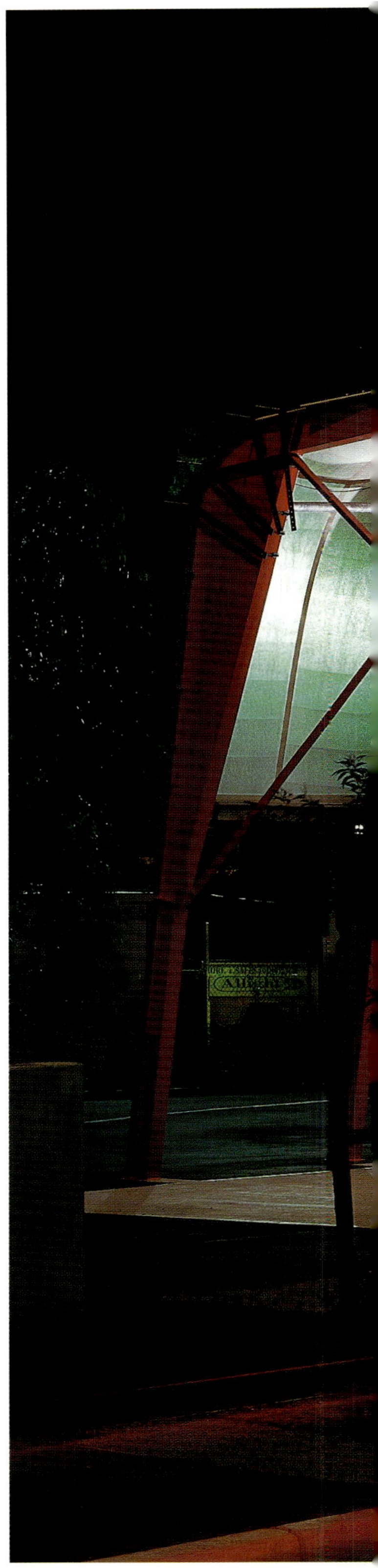

The sites for the amenities buildings are vast planes of tarmac, delineated only by parking bays and road markers. Lightly placed here are the car and coach parking facilities for the Olympic venues. Undaunted by the scale of the surrounding buildings and vehicles, the amenities blocks make their presence felt as a provocation of form and colour, like roadworkers' dayglo vests.

With their primary colours and a series of distorting steel portals swelling under a translucent fabric skin, the buildings act as orientation points for visitors in this vast site, and colour code the various parking facilities. Small urban functions are condensed under their roofs and arcades – queuing, water fountains, telephones and pay stations By day, they are light filled and spacious, By night, they invert, as the fabric glows and magnifies their presence – taut luminescent islands in eddying flows of spectators.

PROJECT TEAM Durbach Block Murcutt; Neil Durbach, Camilla Block, Nick Murcutt, David Jaggers, Lisa Le Van, Joseph Grech

Date of completion 1999

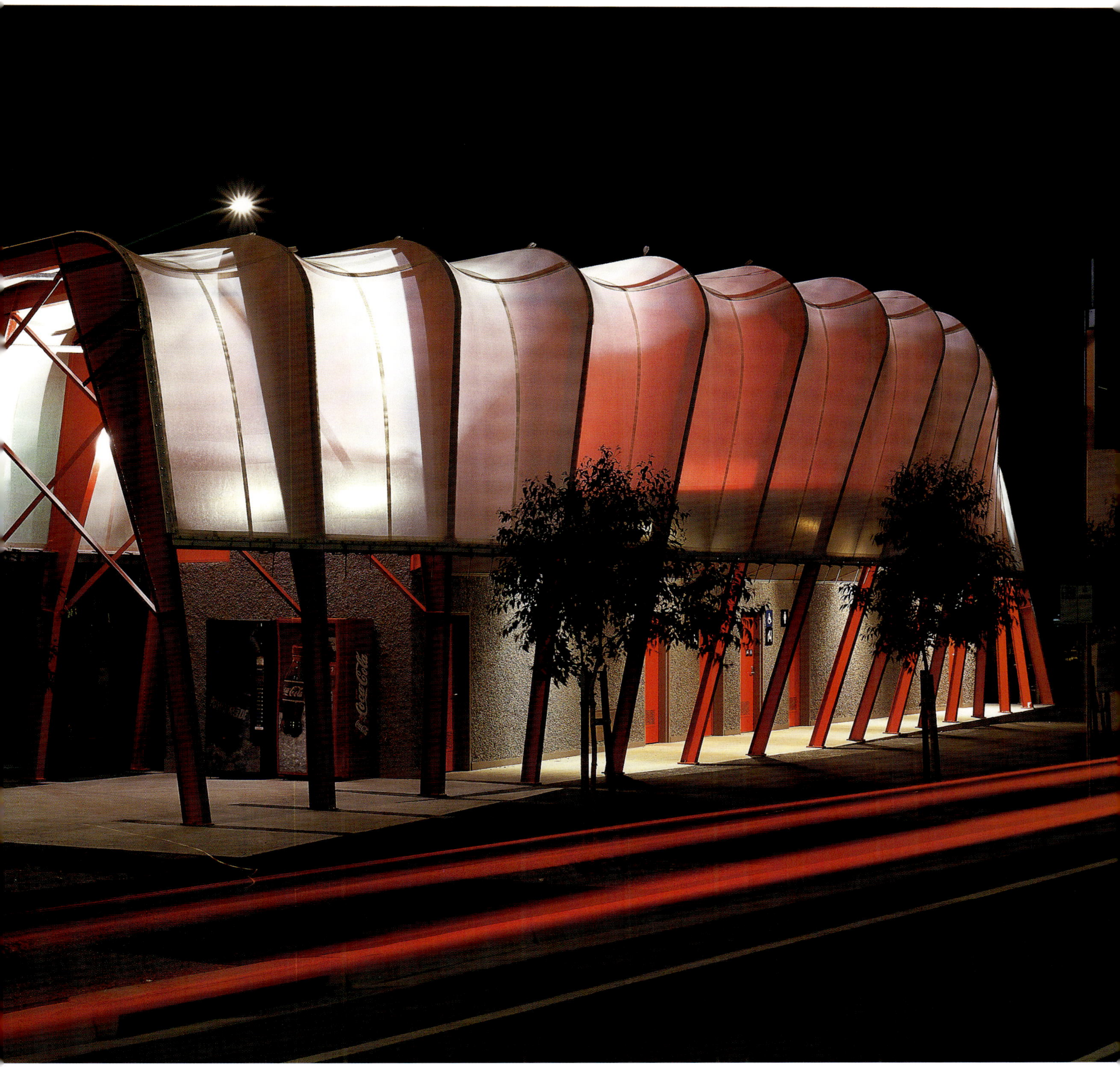
Coca-Cola

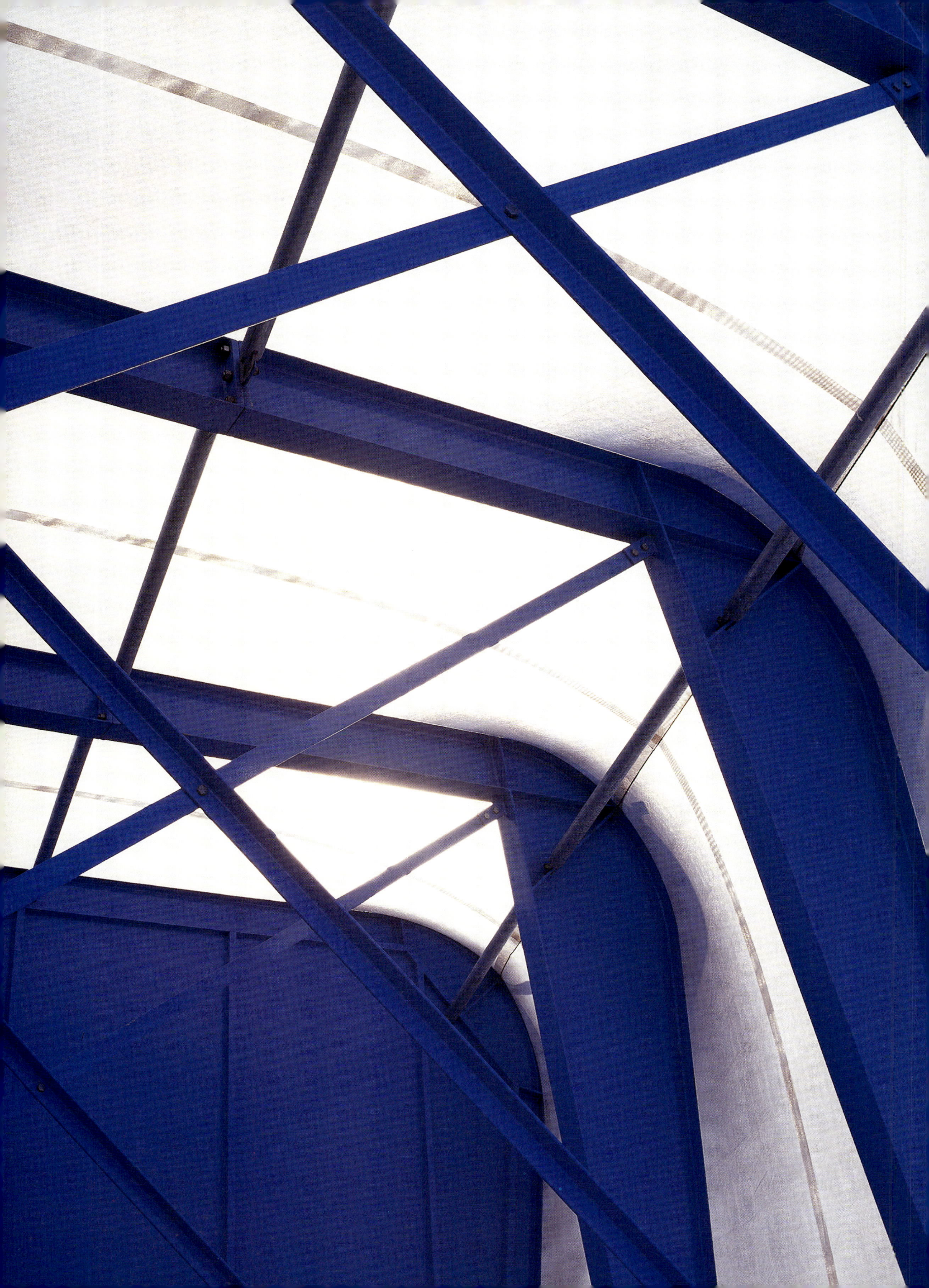

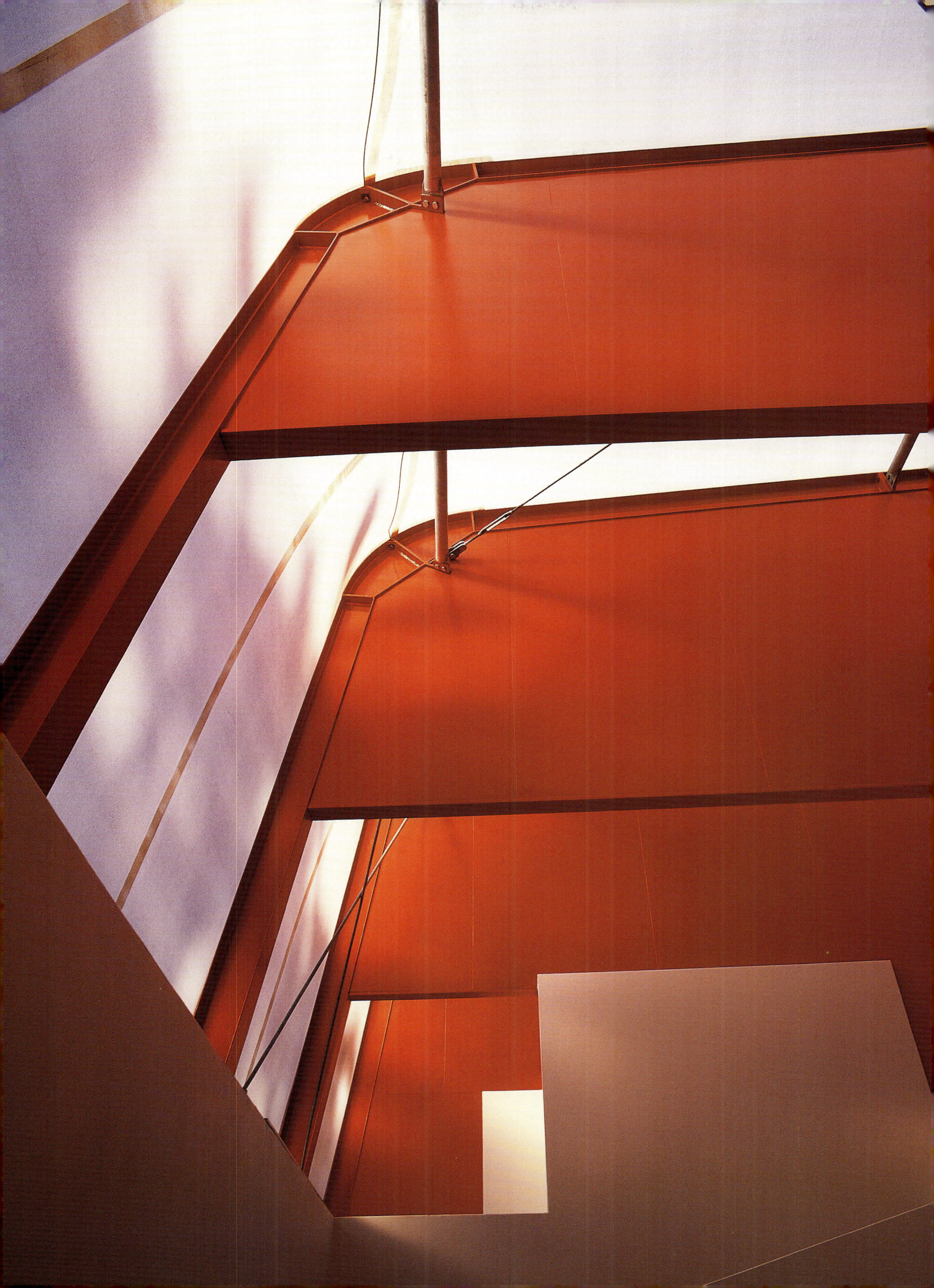

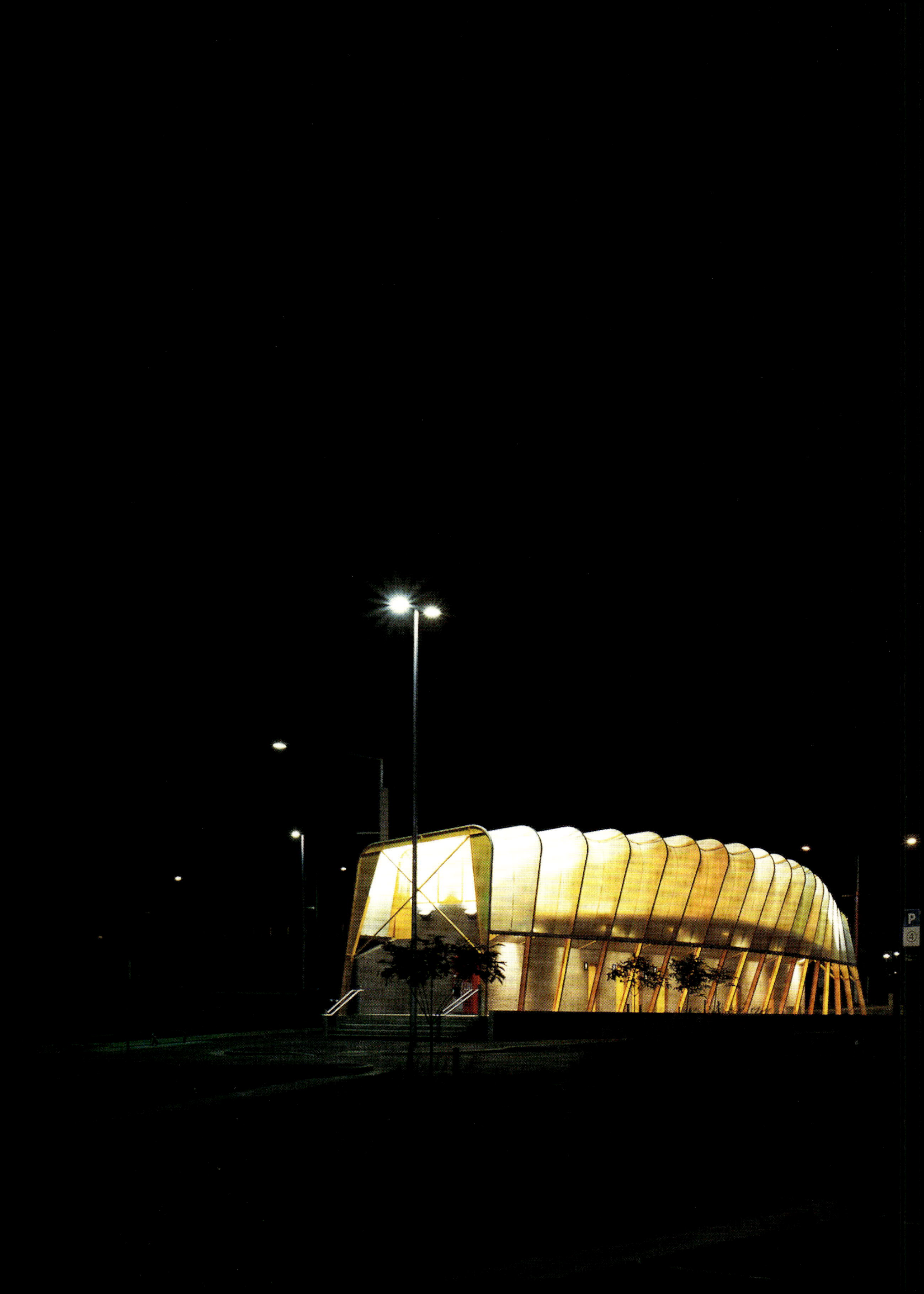

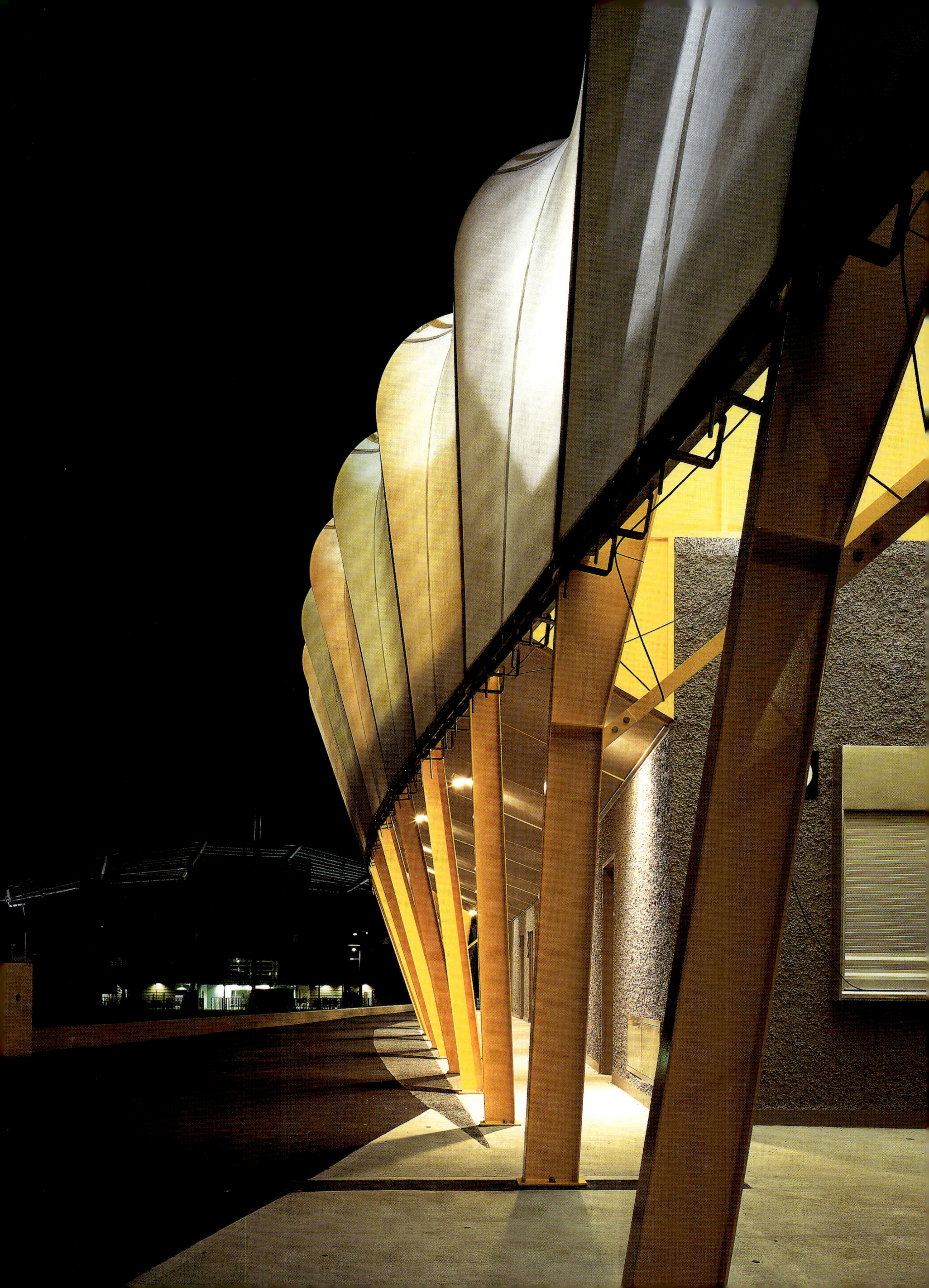

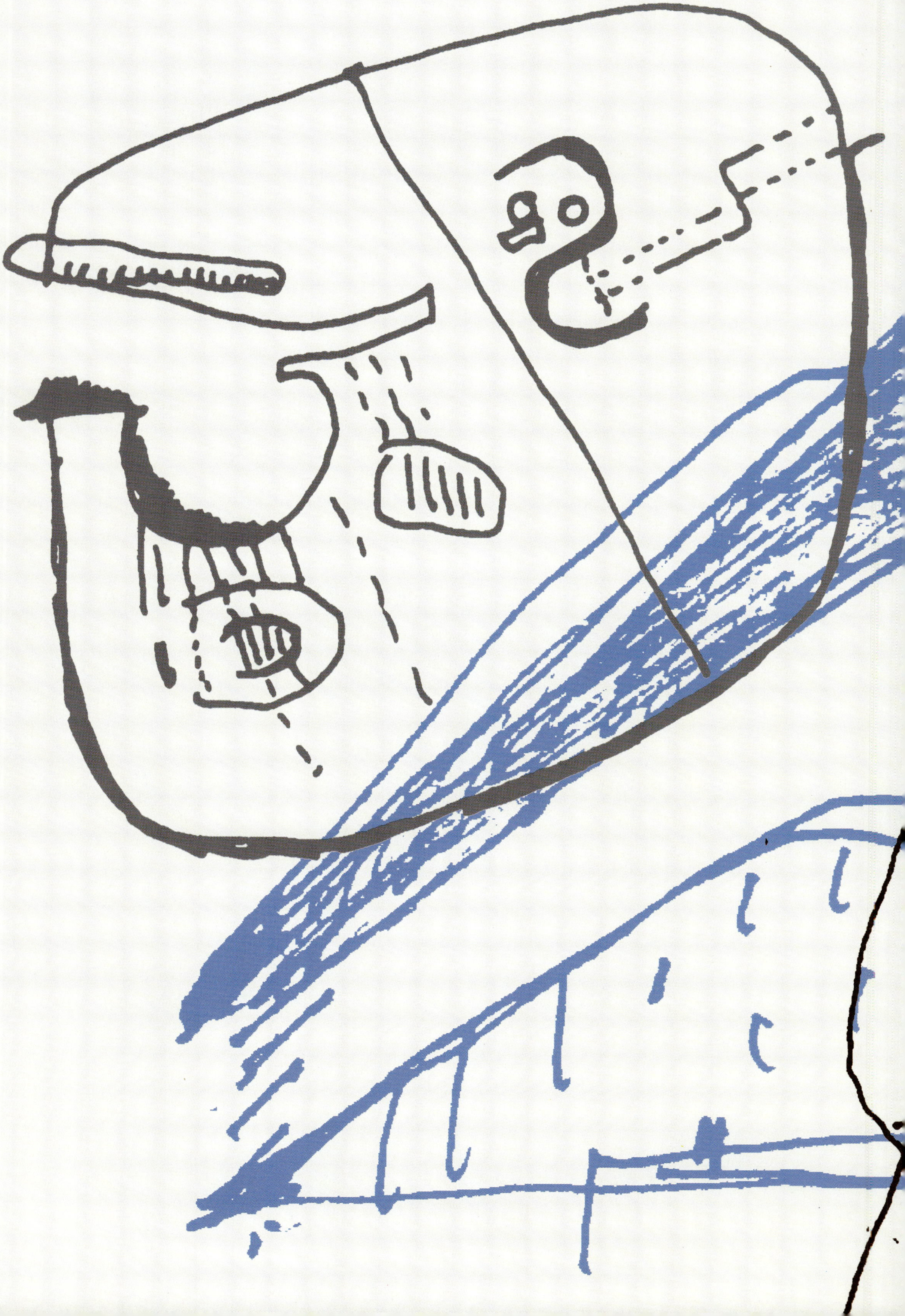

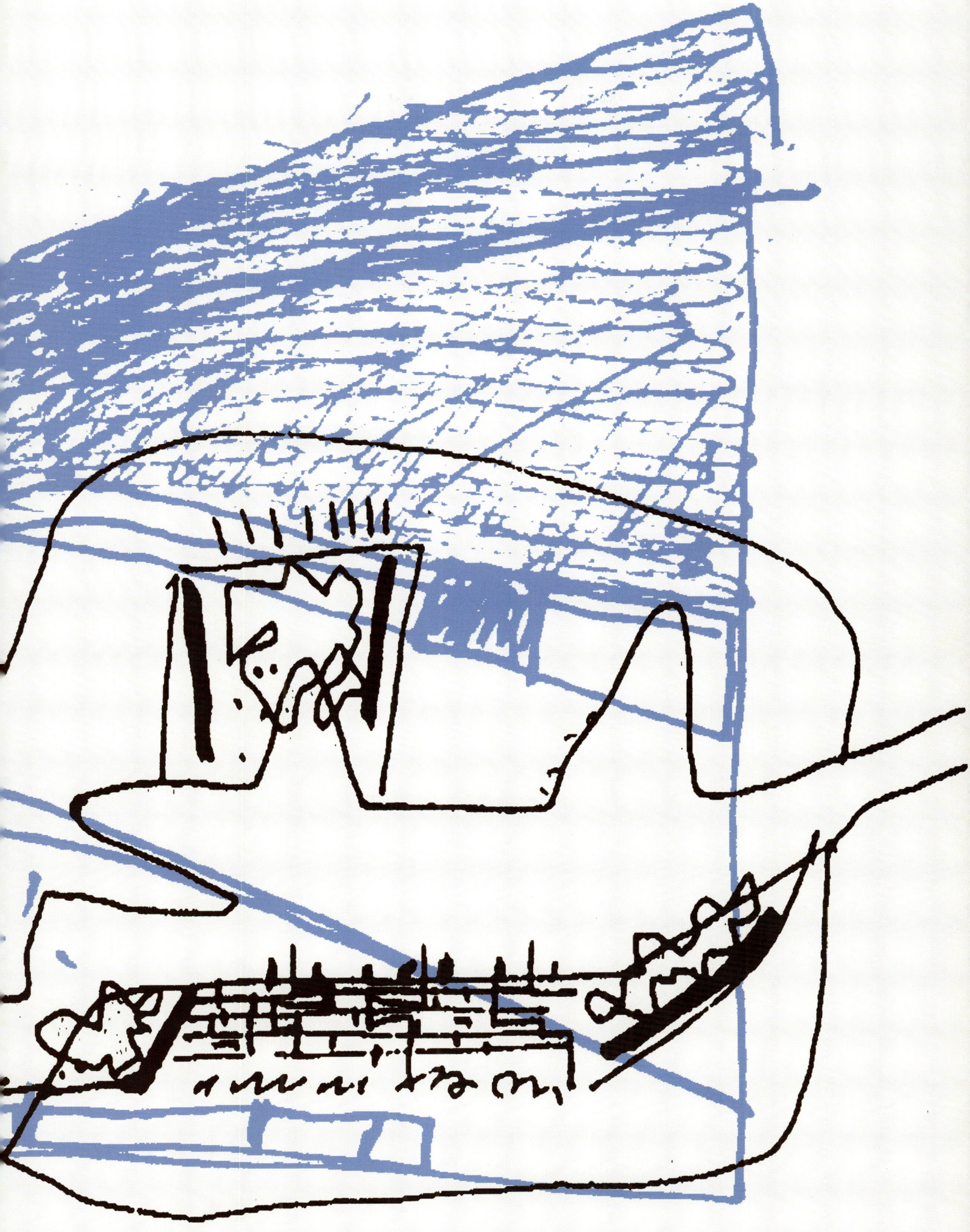

COLLABORATORS

Barnsley, Sue
Block, Camilla
Buchanan, Andrew
Burges, Andrew
Chong, Charles
Collier, Stephen
Cortese, Andrew
Creagh, Lucy
Day, Matt
Durbach, Neil
Grech, Joseph
Jaggers, David
Koopman, Jim
Lassen, Catherine
Le Van, Lisa
Lim, Nancy
Maas, Jeff
Margalit, Harry
Marsh, Rowena
McGrath, James
McInnes, Robin
Metz, Christopher
Murcutt, Nicholas
Murphy, Karen
Ponnambalam, Mano
Ostinga, David
Perlman, Ian
Schaffer, Barbara
Sitta, Tom
Straatveit, Harald
Tesdorf, Nic
Titmuss, Peter
Williamson, Emma
Wilson, Andy

SELECTED WORKS

1999

› House Parker Phillip, Rushcutters Bay
› House Hallen Evans, Mosman
› Apartment Jones, Elizabeth Bay
› House Dewar Leslie, Mosman
› RMIT Student Housing, RMIT Technology Estate, Melbourne
› Invited Entrant – Federation Garden, Centennial Park Corporation, with Schaffer Barnsley and Janet Laurence
› Invited Entrant – Darling Island Housing, Sydney Harbour Foreshore Authority

1998

› Apartment Lowy, Darling Point
› House Sonnabend, Bellevue Hill
› House Levien, Vaucluse
› House Schutzinger, Watsons Bay
› Building 3, Sutherland, Thirty Eight apartments, Community Hall, Retail and parking for GBS Group with Joseph Grech
› Apartments Parramatta Rd, Camperdown, fifty apartments, retail and parking
› ESD Siting and design of Ninety Two units, Illabunda, Winston Hills
› House Levisohn, Greenwich, with Joseph Grech
› Amenities Buildings for the Olympic Coordination Authority, Durbach Block Murcutt
› Rocks Gateway for Sydney Cove Authority with Schaffer Barnsley
› First Place – RMIT Student Housing, RMIT Technology Estate, Melbourne
› Andrew Boy Charlton Pool Competition, East Sydney

1997

› House Hansen, Bellevue Hill
› House Levien, Vaucluse
› The Dig Site, Group Housing and Museum, The Rocks, with Gordon + Vallich, Sydney Cove Redevelopment Authority
› Five Star Hotel and Conference Facility, North Arm Cove for APPM/Goodman Hardie Group, Durbach Block Murcutt
› New Office, Woollahra, for Simons, Nathanson and Marks
› Information Pavilion, Town Hall House, for Sydney City Council
› Temporary Pedestrian Access, Exhibition Centre LRT Stop, for Sydney City Council
› Mining Museum Competition, Kalgoorlie, for Institute of Mining
› Giba Park, Pyrmont Invited Competition for Public Park, City West Development Corporation with Schaffer Barnsley
› Federation Square International Competition
› Public Square, Wintergarden and Cultural Facilities, Melbourne City Centre, Durbach Block Lassen

1996

› Hunters Hill House, Hunters Hill
› House Harding, Jaspers Creek
› Foster Street Warehouse Conversion to Residential Apartments and New Penthouse
› Eco-Tourism Development Communal Facilities, North Arm Cove, for APPM/Goodman Hardie Group, Durbach Block Murcutt
› Cape Barren Island, Tasmania, for APPM/

Goodman Hardie Group, Durbach Block Murcutt
› Development of Self Sufficient Housing for APPM/Goodman Hardie Group, Durbach Block Murcutt

1995
› Apartment Einfeld, Darling Point
› Hillside House, Vaucluse
› House Marincowitz, Clovelly
› Chalmers Street Conversion of existing warehouse to 72 Residential Apartments, Surry Hills
› Master Plan and Building Design, Lake Pillans Wetlands for Lithgow City Council
› Urban Design Proposal, Pyrmont Point, for City West Development Corporation
› Finalist – Civic Centre International Competition, Wagga Wagga City Council, Durbach Block Murcutt

1994
› House Bermeister, Bellevue Hill
› House Vidor & Staub, Watsons Bay
› House Ruch, Byron Bay
› Medium Density Housing, Hobartville, for Windsor Council, Durbach Block Murcutt
› Cardiff Bay Opera House Competition
› Invited participant – Sydney Olympic International Design Workshop

1993
› House Randall, Double Bay
› Darling Point House, Darling Point
› Sketch Design, Pentimento Bookshop, Reading Room and Residence, Balmain
› Masterplan, Chevalier College Bowral
› San Francisco Waterfront Ideas Competition
› Second Place – Honeysuckle Housing Ideas Competition, Newcastle
› First Place – Department of Planning Model Urban Housing Competition, St Clair, Durbach Block Murcutt

1992
› House Grayce, Bondi
› House Rabin, Rose Bay
› Consulting Rooms, Braids, Annandale
› House Vinh, Strathfield, with Harald Straatveit
› House Kemp, Paddington
› Invited entrant – Chifley Square Urban Design Competition
› Student Prize – 'The Next Moves' Medium Density Housing Burwood

1990
› National 'Architects design Furniture' exhibition

1989
› Cumberland Development for Lewis Developments, with Harold Levine
› Antarctic Centre, Sullivan's Cove
› Second and Third stages, First Government House Competition with JTCW

1988
› House Wilson, Woollahra
› New Headquarters for the Royal Australian Institute of Architects (NSW) Tusculum, Potts Point, Levine + Durbach
› Invited Competition – Main Street and Mall, Cronulla, with Nic Tesdorf

1987
› House Packer, Bellevue Hill, with Jim Koopman
› Burwood Council Chambers, Burwood, with JTCW
› Preferred Scheme Invited Entrant – Rocks Square Competition, Levine + Durbach

1986
› House Bryan, Surry Hills
› Joint First Place – South Sydney Ideas Competition, Levine + Durbach

1985
› First Place – New Headquarters for the Royal Australian Institute of Architects (NSW), Levine + Durbach

1984
› Joint First Place – Overseas Passenger Terminal Ideas Competition, Levine + Durbach
› Second Place – Australian Bicentennial Travelling Exhibition, with Richard Johnson and Paul Berkemeier

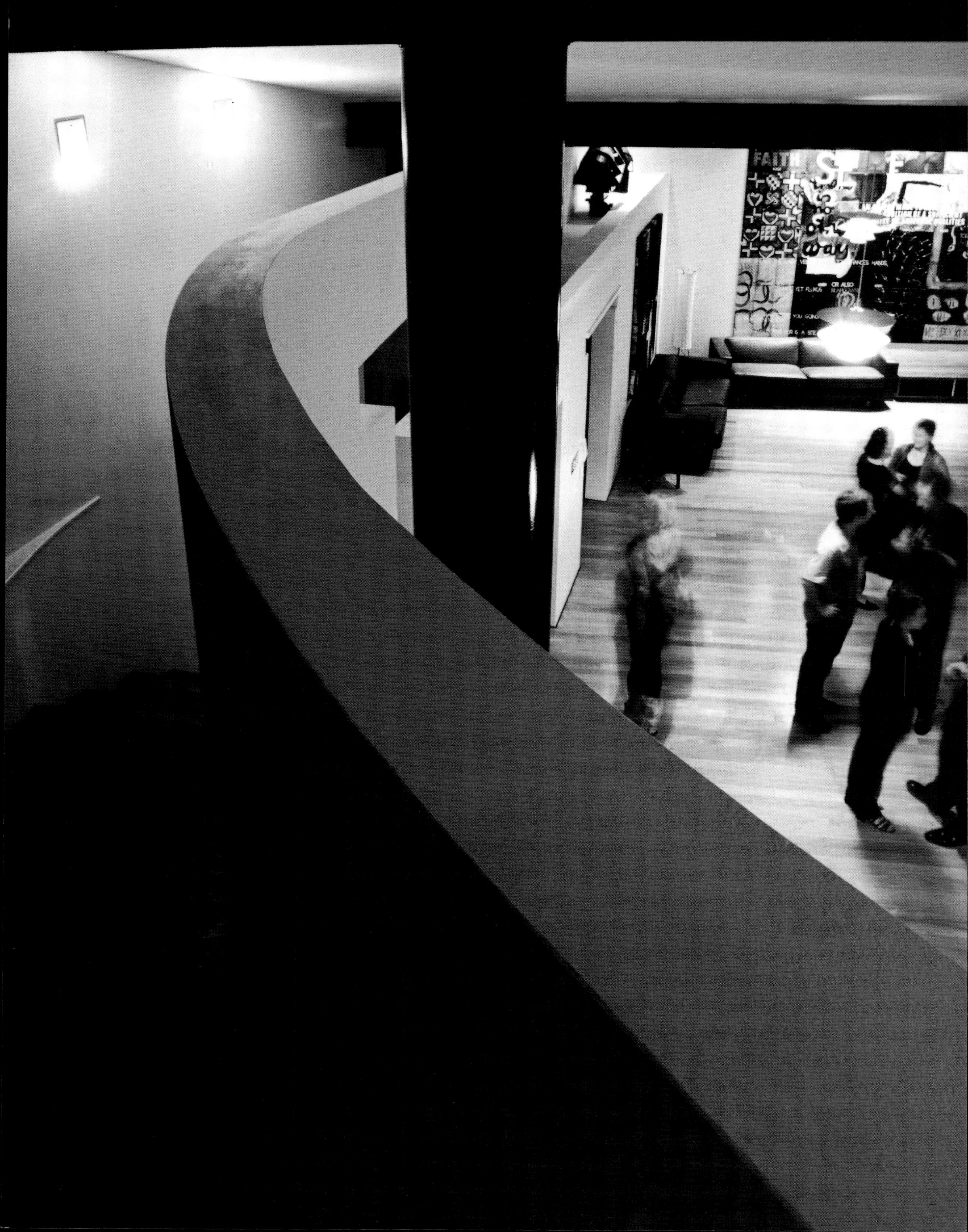
FAITH
way
HANDS.
OR ALSO
YET FLUXUS

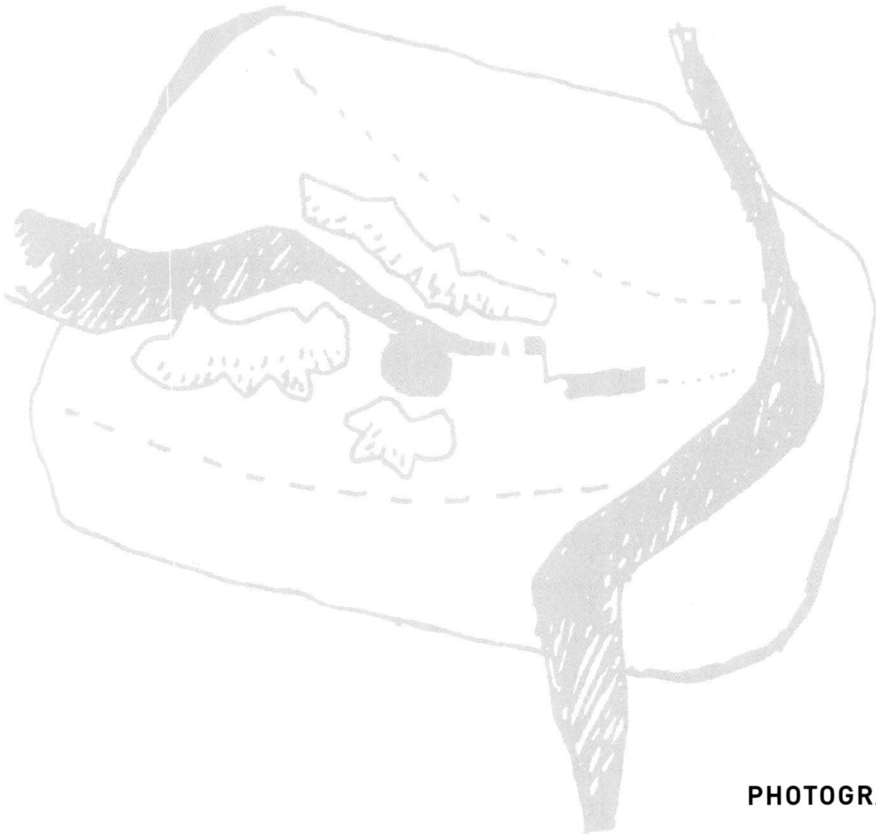

PHOTOGRAPHY CREDITS

All photography © by Patrick Bingham-Hall
except those listed below:

Courtesy of **Durbach Block**
p8, 9, 10, 11, 12, 13, 64-65

John Gollings
p30 top left, 33 bottom

Giorgio Possetti
p30 centre left, 40 top, 41 top

Tim Linkins
p45, 46, 47 top right, centre left,
bottom right, 52 bottom

Eric Sierins
p40 bottom right, 41 bottom

James McGrath
(computer renderings) p76 top left, 82-83